# Mushrooms and Other Fungi

OF THE BLACK HILLS AND SURROUNDING AREA

*Audrey Gabel and Elaine Ebbert*

## Disclaimer

The authors, project staff and Black Hills State University are not liable for or responsible for any identification of any specimen made by users of this book or for the consumption or resulting complications from eating any wild mushroom or fungus. Although some species can be eaten by most people, these species may cause allergic reactions or illness for some people and these reactions cannot be predicted.

*Project Staff:*
Art Director: Linn Nelson
Photography Editor: Steve Babbitt
Business Manager: Dan Farrington
Reviewers: Doug Backlund, Cathy Cripps
Bob Paulson, and Bob Pinette
Illustrators: Kristie Lovett and Larry Ebbert
Scanning: Mark Norby
Production Artist: Kelly Lamb

© 2004 Black Hills State University Press, Spearfish, South Dakota 57799

All rights reserved

Manufactured in the United States of America

No part of this book may be reproduced in any form or by any electronic or mechanical means, including information storage and retrieval systems, without written permission from the publisher, except for brief passages quoted in review.

Cataloging-in-Publication Data

Gabel, A. (Audrey Coxbill)
    Mushrooms and other fungi of the Black Hills and Surrounding Area/A. C. Gabel, E. A. Ebbert.
    p. cm.
    Includes bibliographical references and index.

1. Mushrooms—Black Hills (S.D. and Wyo.)—Identification.
2. Fungi—Black Hills (S.D. and Wyo.)—Identification.
I. Ebbert, E. A. (Elaine Meyer). II. Title. III. Title: Mushrooms & other fungi of the Black Hills and surrounding area.
QK617.G33    2004
579.5'097839

ISBN 0-913062-32-4

Cover Photo: *Boletus edulis* taken by Elaine Ebbert.

*This book was made possible by
the generous contributions of our sponsors.*

Black Hills State University Press

Black Hills State University Bookstore

Black Hills State University Grants and Special Projects

South Dakota Game, Fish and Parks

BLM

Saving the Last Great Places on Earth

National Park Service - Northern Great Plains Network

USDA, USFS Black Hills National Forest

Fred Heidrich, Doug Backlund, Bob Paulson

# Contents

Acknowledgments ... 1
Preface ... 2
How and Where to Collect ... 5
Introduction to Fungi ... 11
Role of Fungi in the Ecosystem ... 12
How to Use this Book ... 13
Major Fungal Groups ... 16
    1. Mushrooms ... 16
    2. Boletes ... 20
    3. Polypores, Toothed Fungi, Jelly Fungi, Ridged Fungi,
       Resupinate Fungi, Coral Fungi and Club Fungi ... 21
    4. Puffballs, Earth Stars, Stinkhorns
       and Bird's Nest Fungi ... 23
    5. Cup Fungi ... 24
    6. Parasitic Fungi ... 25
    7. Lichens ... 26
    8. Slime Molds ... 26
Keys to Mushrooms ... 27
    Non-Brittle, White-Spored Mushrooms ... 31
    Brittle, White to Yellow-Spored Mushrooms ... 49
    Pink-Spored Mushrooms ... 57
    Black-Spored Mushrooms ... 59
    Purple Brown-Spored Mushrooms ... 64
    Orange Brown, Rusty Brown
       to Brown-Spored Mushrooms ... 70
Keys to Boletes ... 76
Keys to Polypores, Toothed Fungi, Jelly Fungi, ... 81
    Ridged Fungi, Resupinate Fungi,
    Coral and Club Fungi
Keys to Puffballs, Earth Stars, Stinkhorns ... 104
    and Bird's Nest Fungi
Keys to Cup Fungi ... 117
Parasitic Fungi ... 127
Lichens ... 132
Slime Molds ... 137
Mushroom Calendar ... 142
Glossary ... 143
Common and Scientific Names of Plants ... 149
References ... 150
Photograph and Illustration Credits ... 152
County Locations of Featured Fungi ... 153
Index ... 159

# Acknowledgments

A very special thanks goes to the South Dakota Game, Fish and Parks Wildlife Diversity Small Grants, for generously providing financial support for four summers of collecting, identifying and photographing fungi. We value the support of Black Hills State University, where the authors were a biology professor and biology student, respectively. A Faculty Research Grant, laboratory and herbarium space, and equipment were provided by the university during a major part of this work.

We wish to extend our appreciation to Linn Nelson, Black Hills State University Assistant Professor and Art Director for her creative talents and her dedication to the scientific as well as the artistic quality of the book. We thank Steve Babbitt, Photography Associate Professor for sharing his expertise and advice regarding quality of photographs and their reproduction. Dan Farrington, Grants and Special Projects Director at Black Hills State University provided exceptional leadership and business skills to move the project to completion.

The valuable suggestions, corrections and numerous contributions of our reviewers, Doug Backlund, Cathy Cripps, Bob Paulson, and Bob Pinette are greatly appreciated. Any mistakes in this book are the fault of the authors and not of our reviewers. A special thanks goes to Kristie Lovett for her outstanding line drawings. Mark Norby, photography student, scanned numerous slides in preparation for the layout and Kelly Lamb worked tirelessly on design and production. We appreciate Larry Ebbert's efforts in creating the map of the Black Hills.

We thank Fred Heidrich, who was so generous with his time and expertise in assisting the writing of a market plan for the book.

The dedication of Kristie Lovett, Sarah Herrin and Steve Mullen, Black Hills State University biology majors, who participated in collecting and identifying fungi during the study is appreciated. We thank the Nelson Endowment for providing summer research scholarships to these students.

Finally, we would like to thank our spouses, Mark Gabel and Larry Ebbert, for their encouragement and constant support during the field work and book preparation.

# Preface

Fungi are fascinating but little known organisms. Despite the lack of knowledge by many people, interest in fungi occurs around the world. Some individuals are interested in fungi because they like to eat mushrooms and they want to know which species are edible, which species are poisonous or which species are hallucinogenic. A fascinating and rich folklore centers around many fungi including those that form "fairy rings". Other individuals enjoy trekking through the forest to collect and admire the handsome and colorful specimens. Some people appreciate their aesthetic value and enjoy photographing elegant specimens. Other individuals are interested in their scientific value such as their role in forest ecology.

## Black Hills

The Black Hills is an eroded, elliptical-shaped dome, located along the western border of South Dakota and northeastern Wyoming, covering approximately 110 miles north to south and 70 miles east to west (FIGURE 1). Annual moisture ranges from less than 16 inches near the periphery to 28 inches in the northern Lead/Deadwood area. The majority of moisture falls from April to September, but late summer is usually dry. Harney Peak, the highest peak is over 7000 ft. and the surrounding plains are approximately 3000 ft. in elevation. Average annual temperatures range from 47º F at lower elevations to 36º F at higher elevations. Chinook winds can quickly change temperatures. The area is described as a unique "island in the plains" where several biomes meet and overlap. The flora includes representatives from the eastern deciduous forest, Rocky Mountains, boreal forests, prairies and the southwest United States. The Black Hills have been described as a distinct ecoregion. Vegetation is diverse, but ponderosa

### Fig. 1  Black Hills of South Dakota and Wyoming

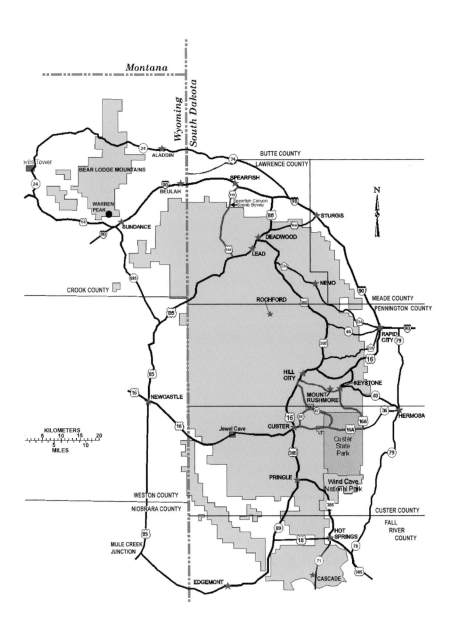

pine (*Pinus ponderosa*) dominates most areas. It is believed that the early Wisconsin glaciation extended into central South Dakota leaving remnants of eastern deciduous and boreal flora in the Black Hills. The moist and cooler climate during glacial periods permitted Rocky Mountain alpine and boreal vegetation to encroach at lower altitudes and latitudes.

Although information exists for plants and animals that occur in the Black Hills, there is almost no literature on the fungi that occur in this region. Because of the need to document the fungi that occur in the Hills, a survey was initiated the summer of 1998. Fungi were collected and identified from numerous sites. The authors were interested in learning where fungal species grew and plant communities associated with different fungi. We wanted to identify species which were commonly collected and species which were uncommon or rare. We were interested in the effect of moisture on their growth and distribution. Information on species occurrence, distribution and ecology are important as a basis to evaluate changes that occur in an area. Such information allows us to identify unique habitats that are rich in fungal diversity and document habitat change and the effect of human activities. This study is on-going and will continue for several years.

The survey was supported by annual grants from the South Dakota Game, Fish and Parks Wildlife Diversity Small Grants. Information from the survey has been included in the South Dakota Natural Heritage Database.

This field guide is an outgrowth of the fungal survey. The fungi included in this book are the more common or interesting fungi that were collected, identified and photographed in the Black Hills. This field guide is unique because it focuses on a specific region, which is the Black Hills and surrounding area. Most fungal field guides cover much larger areas and consequently include fungi that may not be present in a specific region. The book does not include all the fungi that occur in the Black Hills. For example, it does not include some fungi that require microscopic examination for identification or substantial knowledge of mycology (study of fungi) to interpret characters. The book is not a technical book on fungi. The authors have deliberately omitted technical and scientific terms, and avoided using microscopic characters for the identification of fungi included in this book to permit a wide range of readers to be able to use and enjoy this guide. Keys for identification are designed to identify only species included in this book. They cannot be used to identify all the possible species that occur in the Black Hills.

Edibility is based on recommendations of the references used to identify each specimen and these references appear at the end of the book. Since the authors have not evaluated each species for edibility, we recommend taking a very cautious approach when eating fungi. Individuals interested in eating fungi should be very certain their identification is correct. The authors advise anyone planning to eat mushrooms, or any fungus to check their species identification with a professional. Some mushrooms are difficult to identify and many times microscopic characters are the only method to validate a species identification. **If there is any doubt as to the species, do not eat it! You should always test a small amount of a mushroom that you have not eaten before especially if you have food allergies.** There are some deadly poisonous mushrooms in the Black Hills and there are many species that cause gastrointestinal symptoms. A wide range of gastronomic tolerance exists in the population. Some individuals cannot tolerate morels, which are considered safe to eat for most of the population. The edible qualities of many fungi have not been determined and for many the texture or small size of the fungus precludes consideration.

## How to Collect

Individuals who collect fungi usually enjoy and appreciate spending time outdoors and have an interest in fungi. Collecting fungi is an activity that requires patience and determination. Since fungi are very unpredictable as to when they produce fruiting bodies (the structure obvious to the observer), individuals must be willing to search several times and at several locations during the growing season to be rewarded with spectacular collections. Some Black Hills summer days can be dry, sunny and cool, providing an exceptional collecting experience. Other days can be uncomfortable.

The authors can speak from experience after spending much of several summers outdoors searching for fungi. We have experienced snow, sleet, hail, rain, wind and extremely hot temperatures. We have been willing to crawl under logs, climb over downed trees, slide down embankments, crawl through dense vegetation, spend lots of time on our hands and knees and bend over again and again to pick specimens. We have been willing to lie down in mud, manure, ants, poison ivy and other vegetation to photograph specimens.

Other than interest and desire, collecting fungi has few requirements. You can use paper sacks, cloth bags, back packs or plastic bags to carry specimens. If you are interested in eating some specimens, it is best to separate each kind. Wax paper has been used to wrap specimens or they can be placed in separate containers. A hand lens can be helpful to identify field characters and is small and easy to carry in the field. It is important to remove specimens from containers soon after collecting and refrigerate them to retard spoilage. Quality of most mushrooms decreases quickly after picking. Characters for identification such as surface features, color, size, odor and presence of an annulus can be lost quickly so identify them as soon as possible. Field information to record should include location, habitat, substrate and date collected.

Do not collect more specimens than you need or can use. If a very few specimens are present you may choose not to collect at all.

# Where to Collect

Mushrooms and other fungi are where you find them. There is no good way to predict where a species may be growing. Part of the attraction of collecting fungi is the unpredictability of fruiting. However, some general information on habitats can be useful. Puffballs and some mushrooms can be collected from the prairie, or a grassy area, but usually a wooded, wet habitat will be more productive. Since many mushrooms have mycorrhizal (see glossary) relationships with species of trees or groups of trees you would expect to find the fungus only where the tree grows. Timing your collecting a few days after a rain usually improves the productivity. **FIGURES 2-7** show some of the wooded areas where we collected in the Black Hills. Dominant vegetation is listed for each.

The many entities that own and manage land in the Black Hills each have varying regulations regarding collecting. It is imperative that you know the rules pertaining to collecting and trespass for the land you are on before taking anything more than a photograph.

**FIG. 2** A relatively dry east-facing slope in the southern hills dominated by ponderosa pine and Rocky Mountain juniper. Prairie or grassland vegetation is intermixed with the trees.

**FIG. 3** A wet habitat in the central hills exhibiting Black Hills spruce, bog birch, willow and *Sphagnum* moss.

**Fig. 4** A northern hills habitat with trembling aspen, bracken fern and ponderosa pine.

**Fig. 5** A north-facing slope in the central hills dominated by Black Hills spruce, ponderosa pine, trembling aspen and paper birch.

FIG. 6  A wooded canyon in the northern hills with Black Hills spruce, ponderosa pine, trembling aspen, ironwood, bur oak, paper birch, and a small stream.

FIG. 7  A narrow, permanently damp canyon diverging from Spearfish Canyon with ponderosa pine, Black Hills spruce, ironwood, trembling aspen, and paper birch.

# Introduction to Fungi

Fungi typically grow in moist areas. They do not need light and either feed on dead organic material as saprobes, feed on living plants as parasites, or form a relationship with plants in which both the fungus and plant benefit (mycorrhizae). Fungi have vegetative and reproductive stages in their life history. The vegetative stage consists of fine, usually white filaments called hyphae or mycelium. This stage is rarely seen in the field because it is primarily in the soil, in decomposing wood, or in the host plant. A fungus can grow for several years as mycelium and it is considered the perennial portion of the fungus. Recent DNA studies have determined that one of the largest organisms on earth is a fungus. This was documented by sampling mycelium in the soil from an extensive wooded area and finding that the DNA sequences showed all samples were from the same organism, a species of *Armillaria* (Kendrick, 2000).

Only when environmental conditions are favorable does the fungus produce a fruiting body or the sexually reproductive stage from the mycelium. Mycologists have tried to determine the specific environmental conditions for fruiting of a species, but for the most part these conditions continue to elude us. The fruiting body is the part that is visible with the naked eye. Microscopic cells called spores are produced on or in the fruiting body, then released and dispersed. These microscopic spores germinate to form hyphae when environmental conditions are favorable and continue to grow to form masses of mycelium. Spores are dispersed by various mechanisms from specific structures on the fruiting body and these structures will be described for some of the groups discussed. Mushrooms are a type of fruiting body, but other species of fungi produce very different looking fruiting bodies. We describe and illustrate the different types of fruiting bodies for the major groups of fungi that are included in this book.

# Role of Fungi in the Ecosystem

Fungi play several very important roles in the ecosystem. Fungi play an important role as decomposers which is often overlooked. Fungi along with bacteria are the major decomposers of dead plants and animals. They decay or decompose these organic substrates for energy and while performing this process return minerals such as nitrogen, potassium and phosphorus necessary for plant growth to the soil. Because the dead plants on the forest floor are the substrates for this recycling, removal of these substrates ultimately robs them of minerals for growth. Without human interference non-woody and woody plants would remain to be recycled. An equally important role of most forest fungi is their mutualistic relationship with the roots of plants and especially trees. Fungi colonize the roots of pines, spruce, other conifers and deciduous trees to establish a mycorrhizal relationship. The fungi secure their carbon source from the plant (food) and increase absorption of vital minerals (nitrogen, phosphorus and potassium) for the plant.

*Russula* leftovers

This increased absorption is because fungal mycelium extends throughout the soil, beyond roots and increases the surface area for absorption.

Some fungi are plant parasites and obtain nutrients from living plants. This relationship usually is detrimental to the plant that serves as the host. If the host happens to be an

A slug eating *Pholiota terrestris*

economically important plant, attempts such as using fungicides, crop rotation, or breeding for resistance are made to control the parasite and prevent disease. The Black Hills harbors many fungal plant parasites.

Lastly, we cannot ignore the role fungi play in the food chain. Many forest animals feed on fungi. During our years of collection, it was not uncommon to observe mushrooms with bites taken out. We have observed squirrels eating mushrooms. They are selective in which ones they eat. We have observed areas in garages and other outbuildings filled with dried mushrooms, placed there by squirrels. Deer feed on lichens. Small mammals reportedly feed on little hypogeous (underground) fungi (Kendrick, 2000). The animals that feed on mushrooms can be food for raptors and other animals in the food chain.

## How to Use this Book

This field guide is unique in that it focuses on mushrooms and other fungi that occur in the Black Hills or adjacent areas. Every specimen featured in this book was collected in this area and photographs are exclusively of Black Hills mushrooms and fungi. Most field guides cover much larger geographical areas and consequently include specimens not occurring in a collector's immediate area.

This book is designed for a variety of readers and can be used by individuals with little or no background in studying fungi. Much of the scientific terminology has been omitted and we have attempted to define terms the first time they are used as well as include them in the glossary. You will not need a microscope to identify fungi in this book.

First look at the eight major group headings on pages 16, 20, 21, 23, 24, 25, 26. Immediately following the first five groups are pictures of representatives in that group. All groups include descriptions of fungi in that group. After looking at the photos and reading the description for each group, try to place your specimen in one of the eight groups. For example, Group 1 on page 16 is the group for Mushrooms. Knowing you have a mushroom you can either go to the Key to Mushrooms (page 27) to help you identify your specimen or page through the pictures and descriptions of mushrooms (pages 31–75 ) to locate one that looks like your specimen. An explanation on how to use a key is on page 15.

You should notice that mushrooms are divided into groups based on spore color. Spore color is sometimes the color of the gills (lamellae), but after tentative field identification, always make a spore print (See page 18 to learn how to make a spore print.) to be sure. Once you have picked the correct spore color check all the other features such as size, cap shape and color, stalk shape and color, gill attachment, presence or absence of a ring (annulus) and presence or absence of a cup (volva). Range of sizes of fruiting bodies are the ranges observed in the Black Hills during this study. It is possible that elsewhere, under different enviromental conditions, sizes may be larger or smaller. The range of spore sizes usually follows Arora (1986), because spore sizes of our collections fit into that range.

We have included both common and scientific names for each fungus featured. Scientific names are in Latin and have two parts. The first part is the genus and the second part is the specific epithet.

**We have NOT included all the possible fungi that occur in the Black Hills and we have not included fungi that DO**

require microscopic examination for identification. It is very possible that your specimen may not be in this book. If this occurs you need to consult more comprehensive field guides or consult a professional. References of other field guides are listed at the end of this book. DO be sure of your identification, if you are planning to eat them.

## How to Use a Key

A key is a tool that can be used to identify fungi. The key consists of a system of statements usually arranged in pairs (couplets), asking the user to pick one of the statements that describes the specimen in hand and follow the number listed on the right to continue through the key. The Group 1 - Key to Mushrooms on page 27 begins with a set of six statements describing spore color (1A, 1B, 1C, 1D, 1E, 1F). For example, let's say that you have determined that your mushroom agrees with statement 1A, which states that the spore print is white and the mushroom's texture is not brittle. On the right of that statement is the number 2. Skip down to the statements with the number 2 (2A, 2B and 2C). Read each statement and decide which one describes your mushroom. If you decide your mushroom has white, free gills and both a ring and cup you would skip down to the number 3 couplet. From this point on all statements are in couplets. At this point you must decide if your mushroom has a red to orange cap with white patches or if it is not this color, but white to tan and without white patches. If it agrees with 3A you have the Fly Agaric and you have identified your specimen. Now go to the photo and description of the Fly Agaric and determine if you have made the correct choice. If it agrees with 3B you will need to go to couplet 4 and continue through the key until you have identified your specimen.

If after examining the photo and reading the description of your specimen you decide you have not made the correct choice, start at the beginning of the key and carefully read the statements and observe the features of your specimen again. Keys attempt to pick the important characters and carefully describe each to correctly identify a specimen. However, they are not perfect.

# Major Fungal Groups

## Group 1 - *Mushrooms*
*(Refer to pages 27-75)*

Hygrophorus speciosus

Cortinarius species

Amanita bisporigera group

Clitocybe gibba

**Fig. 8**

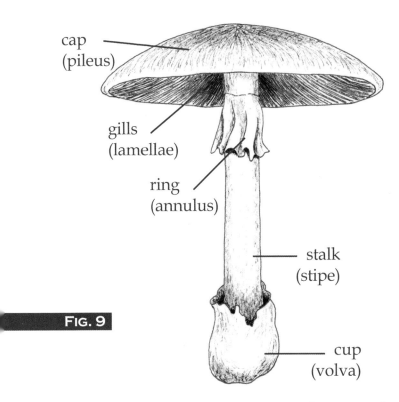

FIG. 9

Mushrooms are fleshy, non-woody fungi **(FIGURE 8)**. A typical mushroom consists of a flat to rounded cap (pileus) on a stalk (stipe). The underside of the cap contains thin plates called gills (lamellae) **(FIGURE 9)**. Spores are produced on special cells (basidia) on these gills and when mature are forcibly released, picked up by air currents and dispersed. Spores can be white, pink, yellow, various shades of brown and shades of grey to black. Gill color can reflect spore color, but this is not always true. Gills can change color as the mushroom matures. In some mushrooms the edge of the cap is attached to the stalk when immature and as the mushroom matures the cap expands breaking away from the stalk, leaving a ring (annulus) around the stalk at maturity **(FIGURE 8,** *Cortinarius* species., and **FIGURE 9)**. Another group of mushrooms is enveloped in an "egg-like" structure when immature and when the cap and stalk emerge from the "egg", a cup (volva) remains at the base of the stalk. Some mushrooms have both an annulus and a volva **(FIGURE 8,** *Amanita bisporigera* **group)** and other mushrooms have neither an annulus nor a volva **(FIGURE 8,**

**17**

*Hygrophorus speciosus* and *Clitocybe gibba*). Mushrooms grow on a variety of substrates that include soil (terrestrial), wood, dung, straw and compost. Some mushrooms grow on other mushrooms. Some mushrooms are edible, but some are deadly poisonous.

**In this book field characters to use for identification include:**
1. Spore color as shown by a spore print and sometimes gill color. Although you can use gill color in the field for tentative identification, always verify spore color with a spore print. To obtain the correct spore color cut off the cap at the top of the stipe and lay it gills down on a piece of paper (white paper if you predict the spores are dark, black paper if you predict white spores or both white and black paper if you cannot predict). Place a bowl or cup over the cap for a few to several hours. The spores will be deposited on the paper to show the color **(FIGURE 10)**.
2. Color surface features, texture and size of cap. It is important to record this data in the field because color and size change as the mushroom dries.
3. Color and size of stalk. Record in the field because color and size change during the drying process.
4. Presence or absence of a cup (volva). You must dig below the soil line to observe cups. Since some very poisonous mushrooms have volvas, you do not want to miss this character.
5. Presence or absence of a ring (annulus). The ring may dry up and appear to be absent. Examine young specimens in the field for evidence of rings.
6. Shape of cap **(FIGURE 11)**.
7. Gill (lamellae) attachment. Some mushrooms have gills that connect to the stalk (attached) at the top and in others the gills terminate before they reach the stalk (free) **(FIGURE 12)**.
8. Record any odor that may be present.
9. Record the substrate upon which the mushroom is growing, ie. soil, wood, dung, straw or compost.

Some mushrooms may need both field and microscopic characters for identification.

**Fig. 10** Spore Print.

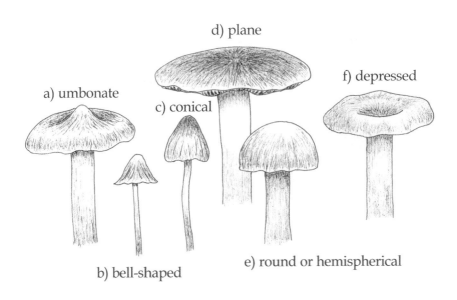

**Fig. 11** Cap Shapes.

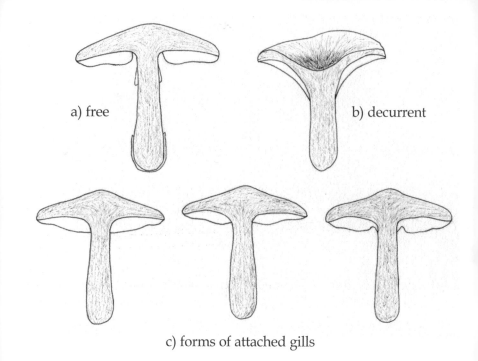

a) free

b) decurrent

c) forms of attached gills

**FIG. 12** Gill Attachments.

## GROUP 2 - *Boletes*
*(Refer to pages 76-80)*

*Suillus granulatus*

**Fig. 13**  *Leccinum insigne*

Boletes **(FIGURE 13)** appear similar to mushrooms except the lower surface of the cap consists of many tiny pores, or tubes in which the spores are produced. Field characteristics similar to those used for mushrooms are needed to identify boletes. Boletes usually do not have cups and nearly always are terrestrial. Some boletes are edible.

## GROUP 3 - *Polypores, Toothed Fungi, Jelly Fungi, Ridged Fungi, Resupinate Fungi, Coral and Club Fungi,*
*(Refer to pages 81-103)*

Fomitopsis pinicola　　Sarcodon imbricatum　　Tremella mesenterica

**21**

Gomphus clavatus    Ramaria apiculata    Clavariadelphus truncatus

**FIG. 14**

As the heading implies, there is a great deal of variation in the appearance of fruiting bodies in this group. Fruiting bodies have either woody, jelly-like, leathery or elastic consistency. Many of these fruiting bodies are persistent and do not decay quickly like mushrooms. They are found growing on dead wood, wood of living trees, and on the soil. Polypores can be mushroom-shaped with a stalk and a cap, but more commonly appear as brackets or shelves **(FIGURE 14,** *Fomitopsis pinicola***)**. As their name indicates, the lower surface consists of many tiny pores and the cells which produce the spores line the inside of these pores. Since they are firm and woody they can be separated easily from boletes. Toothed fungi resemble polypores, except the lower surface bears teeth-like structures around which the spores are formed. Toothed fungi **(FIGURE 14,** *Sarcodon imbricatum***)** can have stalks or a bracket attachment. Because of the tough texture, most of the fungi in this group are not palatable. When wet, jelly fungi **(FIGURE 14,** *Tremella mesenterica***)** have a conspicuous elastic, tough jelly-like texture. Under dry conditions they shrivel to inconspicuous growths on the dead wood. Return of moisture will rehydrate them. The most common species of jelly fungi in the hills are yellow or black. Still others in this group are cornucopia-shaped and spores form on ridges on the lower or outer surface **(FIGURE 14,** *Gomphus clavatus***)**. This group would include the edible chantherelles.

Coral fungi are highly branched and resemble marine corals, which are animals **(FIGURE 14,** *Ramaria apiculata***)**.

A closely related club-shaped group are the club fungi (**FIGURE 14,** *Clavariadelphus truncatus*). Their smooth surface includes spore-producing areas. Another group has flat (resupinate) fruiting bodies.

## GROUP 4 - *Puffballs, Earth Stars, Stinkhorns and Bird's Nest Fungi*
*(Refer to pages 104-116)*

*Lycoperdon perlatum*

*Geastrum quadrifidum*

*Crucibulum laeve*

*Phallus impudicus*

**FIG. 15**

These groups form a category of fungi with spores enclosed in structures. Puffballs are appropriately named because the fruiting body is round and most of them release a cloud of spores when they are handled. Most puffballs have no stalk, no cap, no gills or pores (**FIGURE 15,** *Lycoperdon perlatum*). The spores are contained inside the fruiting body ("ball"). Most of them are soft. Some release the spores through a tiny hole, which forms on the top. Others break down or start falling apart and the spores are released. Giant puffballs sometimes are found in pastures and can reach 1-2 ft. in diameter. Most puffballs are edible prior to spore formation when they are firm and white. There is a less common group of puffballs that are hard and these are never edible. A category of puffballs, called earth stars **(FIGURE 15,** *Geastrum* species**)** possess an outer wall layer that splits into rays and reflexes to give the appearance of a star, on which the spore-containing structure sits.

Bird's nest fungi are very small **(FIGURE 15,** *Crucibulum laeve***)**. They consist of an outer wall (exoperidium) which resembles a "nest" and the "nest" contains egg-like structures (peridioles) in which spores are produced. Most collectors know when they have collected a stinkhorn because they are foul smelling **(FIGURE 15,** *Phallus impudicus***)**. The smell has been compared to rotten meat. They emerge from an "egg" (often buried in the soil). A spongy stalk pushes up a wrinkled cap covered with a dark gelatinous matrix, which contains the spores. The smell attracts flies to the gelatinous matrix and they serve to disperse the spores. Fungi in **GROUP 4** grow on the soil, dead wood and vegetation or less frequently on dung.

## GROUP 5 - *Cup Fungi*
*(Refer to pages 117-127)*

*Gyromitra gigas*

*Peziza repanda*

*Helvella ephippium*

*Morchella angusticeps*

**FIG. 16**

Some fungi form fruiting bodies which are cup-shaped (**FIGURE 16,** *Peziza repanda*) or convoluted (**FIGURE 16,** *Gyromitra gigas, Helvella ephippium*). Some form fruiting bodies with ridges and pits (**FIGURE 16,** *Morchella angusticeps*). In all cases the spores are produced inside of the cup. Morels and false morels are included in this group. Cup fungi are found on soil or wood. Colors include shades of brown, red, orange or yellow.

## GROUP 6 - *Parasitic Fungi*
*(Refer to pages 127-132)*

Included in the field guide are a few examples of easily observed parasitic fungi occurring in the Black Hills. Although there are many species of plant parasitic fungi throughout the Black Hills and the world, most are microscopic and not included in this field guide.

# GROUP 7 - *Lichens*
*(Refer to pages 132-136)*

Lichens are not only fungi, but organisms consisting of both a fungal and algal component. These two groups of organisms exist in a compatible relationship where both benefit. The alga photosynthesizes, providing the sugars for the fungus and the fungus provides a protected habitat and increased access to minerals for the alga. A few lichens have cyanobacteria contributing the sugars.

Lichens do not resemble a fungus or an alga, but develop a totally different growth form called a thallus. Some grow flat and attach very tightly to the substrate and are categorized as crustose lichens. Other are flat and leaf-like, attaching to the substrate, but with free edges. These are foliose lichens. Fruticose lichens either hang from the substrate or produce vertical structures. Lichens commonly grow throughout the Black Hills on soil, moss, trees or rocks. Although you may see them associated with living trees, they are not parasites. A 1968 publication by C. Wetmore, listed in the Selected References focuses on Black Hills lichens.

# GROUP 8 - *Slime Molds*
*(Refer to pages 137-141)*

Another group of organisms that are not fungi, but commonly studied by mycologists, are slime molds. We include this group because they occur in the hills and most are elegantly beautiful. People infrequently observe slime molds because the fruiting bodies are very small, usually less than $\frac{1}{4}$ inch. However, if you are willing to examine punky logs on your hands and knees you might see their tiny beautiful fruiting bodies. These organisms have two very distinct portions to their life history. When moisture abounds, a stage called a plasmodium, which resembles a giant amoeba, crawls and feeds on bacteria decomposing vegetation. Plasmodia can be colorless, white, pink, yellow or other colors. If colored, the plasmodium can be observed as a flat slimy growth on logs or other vegetation. However, the plasmodium is infrequently observed because it is ephemeral and secretive, growing among leaves and under bark. It emerges when it is time to develop fruiting bodies, which contain spores. Spores disperse and fuse with other cells to form the plasmodium again.

## GROUP 1 - *Keys to Mushrooms in Field Guide*

1A. Spore print white; mushroom texture not brittle ..... 2
1B. Spore print white to yellow; mushroom texture brittle ..... 17
1C. Spore print pink ..... 23
1D. Spore print black ..... 24
1E. Spore print purple-brown to gray-brown ..... 26
1F. Spore print orange-brown, rusty-brown to brown ..... 29
2A. Gills white and free; ring present but no cup ..... Smooth Parasol *(Lepiota naucina)* p.31
2B. Gills white and free; both ring and cup present ..... 3
2C. Gills various colors, including white, attached or decurrent; no cup ..... 5
3A. Cap red to orange, with white patches of universal veil remnants attached ..... Fly Agaric *(Amanita muscaria* var. *formosa)* p. 32
3B. Cap white, gray or tan ..... 4
4A. Cap gray or tan, round becoming plane at maturity; ring absent ..... Grisette *(Amanita vaginata* group*)* p. 33
4B. Entire mushroom white; ring and cup prominent ..... Destroying Angel *(Amanita bisporigera* group*)* p. 34
5A. Stalk short, lateral or eccentric; white to tan cap; gills white; growing as bracket on some deciduous trees ..... Oyster Mushroom *(Pleurotus populinus), (Pleurotus ostreatus)* p. 35
5B. Stalk obvious and centrally located ..... 6
6A. Gills distinctly decurrent ..... 7
6B. Gills not distinctly decurrent ..... 9

**7A.** Cap yellow to tan, deeply depressed in the center; white to cream decurrent gills....Funnel Cap *(Clitocybe gibba)* p. 36
**7B.** Entire mushroom gold to orange........8
**8A.** Decurrent gills with connecting veins; cap 1.0 in.; in groups on dead wood............Tiny Trumpets *(Xeromphalina campanella)* p. 37
**8B.** Decurrent gills dichotomously branched, no veins; cap wider than 1.0 in............False Chanterelle *(Hygrophoropsis aurantiaca)* p. 38
**9A.** Gills waxy, distant and attached (sometimes weakly decurrent); cap brightly colored and usually wet and slimy........10
**9B.** Gills not waxy, but attached........12
**10A.** Cap yellow to orange to red, conical; gills attached; mushroom turning dark blue/black when bruised or aged........Conical Waxy Cap *(Hygrocybe conica)* p. 39
**10B.** Not turning blue/black........11
**11A.** Cap yellow to orange to red, wet and slimy; gills weakly decurrent......Pine Waxy Cap *(Hygrophorus speciosus)* p. 40
**11B.** Mushroom white to pale yellow; yellow flakes or granules on cap and stalk............Yellow Waxy Cap *(Hygrophorus chrysodon)* p. 41
**12A.** Gray and fragile; cap conical to convex; cap flesh thin to show gills........Gray Dunce Caps *(Mycena pura)* p. 42
**12B.** Not fragile; cap not conical and cap flesh thicker........13
**13A.** Ring present; cap round becoming plane at maturity............Honey Mushroom *(Armillaria ostoyae)* p. 43
**13B.** Ring absent........14
**14A.** Robust mushroom, white or shades of pink to lavender; plane cap 2-6 in. wide........15
**14B.** Mushrooms another color........16
**15A.** Mushroom pale pink to lavender to blue when fresh............Blewit *(Clitocybe nuda)* p. 45
**15B.** Mushroom white............White Leucopax *(Leucopaxillus albissimus)* p. 46
**16A.** Cap egg-yolk yellow with reddish streaks; gills yellow............Plums and Custard *(Tricholomopsis rutilans)* p. 47
**16B.** Cap tan; gills white; stalk flattened near the top............Oak Gymnopus *(Gymnopus dryophilus)* p. 48

**17A.** Cap tan to gray; stalk at least ¼-½ in. wide; when fresh, gills exuding a latex when cut ... 18
**17B.** Cap various colors; gills not exuding latex when cut ... 19
**18A.** Latex colorless; odor similar to maple syrup; in wet boggy areas
......... Burnt Sugar Mushroom *(Lactarius aquifluus)* p. 49
**18B.** Latex milky; odor not similar to maple syrup; in wet boggy areas
......... Milky Mushroom *(Lactarius mammosus)* p. 51
**19A.** Cap entirely pink or red; spores white to yellow
......... Rosy-Red Russulas *(Russula* species) p. 52
**19B.** Cap color variable, but not entirely pink or red; spores white to yellow ... 20
**20A.** Cap variable with shades of yellow, orange, tan or red; spores yellow
......... Variable Russula *(Russula alutacea* group) p. 53
**20B.** Cap white, tan or gray (may have some shades of pink); spores white ... 21
**21A.** Mushroom tough and hard ... 22
**21B.** Mushroom not tough, but soft; cap white to include overtones of pink or lavender; spores white
......... Fragile Russula *(Russula fragilis)* p. 54
**22A.** Mushroom changing from white to gray to black after picking or at maturity; cap center depressed
......... Blackening Russula *(Russula albonigra)* p. 55
**22B.** Mushroom not changing colors; cap center depressed; very short stalk, just emerging from soil line
......... Short-Stemmed Russula *(Russula brevipes)* p. 56
**23A.** Cap white to tan, smooth with fibrils; stalk white; gills free and pink at maturity
......... Fawn Mushroom *(Pluteus cervinus)* p. 57
**23B.** Cap and gills salmon, surface of cap netted
......... Netted Mushroom *(Rhodotus palmatus)* p. 58
**24A.** Cap white to gray to tan; gills free, black and deliquesing when mature ... 25
**24B.** Cap tan to brown; gills attached, black when mature and not deliquesing
......... Bell-Shaped Panaeolus *(Panaeolus campanulatus)* p. 59

**25A.** Cap white to gray, cylindrically-shaped and shaggy......................Shaggy Mane *(Coprinus comatus)* p. 60

**25B.** Cap white to gray to tan, conical or bell-shaped..........Mica Cap *(Coprinus micaceus)*, plus other species p. 61

**26A.** Cap white to tan; gills free, dark brown at maturity; ring usually present..................Woodland Agaricus....................*(Agaricus silivicola)*, plus other species p. 64

**26B.** Cap tan to more brightly colored; gills attached, dark brown at maturity; with or without a ring..................27

**27A.** Upper surface of cap firm, yellow to tan with yellow-brown scales; stalk with delicate white scales, $\frac{1}{2}$-$\frac{3}{4}$ in. thick; ring present..................Scaly Stropharia *(Stropharia kaufmanii)* p. 66

**27B.** Upper surface of cap soft, gray to tan; stalk < $\frac{1}{2}$ in. wide..................28

**28A.** Cap round to plane; stalk ~ $\frac{1}{4}$ in. Common Woodland..................Psathyrella *(Psathyrella candolleana)* p. 67

**28B.** Cap conical to bell-shaped; stalk $\frac{1}{4}$ in. or less in width..................Graceful Psathyrella *(Psathyrella gracilis)* p. 68

**29A.** Cap yellow to rusty brown; spores more orange than brown; on wood..................Common or Boring..................Gymnopilus *(Gymnopilus sapineus)* p. 70

**29B.** Cap yellow, tan, brown; spores brown to rusty brown...30

**30A.** Cap yellow to brown, changing colors as it dries with periphery darker brown; on dead wood..................Vernal Changing Pholiota *(Pholiota vernalis)* p. 71

**30B.** Cap tan to brown, not dramatically changing color as above; on soil..................31

**31A.** Cap tan-brown, conical to plane, splitting at the margin; silky texture with an umbo; spores dull brown..................Silky Little Brown Mushroom *(Inocybe fastigiata)* p. 72

**31B.** Cap tan-brown, round to plane, sometimes with an umbo; spores rusty brown; cobwebby veil present on immature specimens..................*(Cortinarius* species*)* p. 73

# GROUP 1 - *Mushrooms*

## NON-BRITTLE — WHITE-SPORED MUSHROOMS
### SMOOTH PARASOL - *Lepiota naucina*
### (=*Leucoagaricus naucinus*)

**CAP:** Round to hood-shaped when immature, becoming flat at maturity, white and smooth, 5-9 cm (2-3 ½ in). **GILLS:** White, crowded and free. **STALK:** White, 5-9 cm (2-3 ½ in) X 1-1 ½ cm (¼-½ in), expanding to form an enlarged base, prominent ring present. **SPORES:** White, smooth, 7-9 X 5-6 µm **HABITAT:** Collected in meadows or pastures, adjacent to mixed coniferous/deciduous woods. **EDIBILITY:** It has been listed as edible, but the authors recommend **NOT** eating it because of the similarities with poisonous *Amanita* species. **COMMENTS:** Species in the genus *Lepiota* can be easily confused with species in the genus *Amanita*. *Amanita* species have BOTH a cup and a ring. Species in the genus *Lepiota* have a ring but not a cup. A novice collector might mistake the bulbous base of the stalk of a *Lepiota* for a cup. It is also very easy to miss the presence of a cup if the collector does not dig below the soil line when collecting the mushroom.

## FLY AGARIC - *Amanita muscaria* var. *formosa*

**CAP:** Round becoming flat at maturity, yellow/orange to orange/red with white patches of cup remnants, 8-25 cm (3-10 in). **GILLS:** White, darkening with age, crowded and free. **STALK:** White, 8-10 cm (3-4 in) X 1.5-2 cm (½-¾ in) becoming wider at the base, distinct white ring present. Stalk emerging from a cup of concentric rings,

which may be buried in the soil. **SPORES:** White, smooth, 9-13 X 6.5-9 μm. **HABITAT:** Collected infrequently in mixed coniferous/deciduous woods in the Hills. **EDIBILITY: POISONOUS!** The mushroom contains muscimal and ibotenic acid, which affect the central nervous system and cause delirium. **COMMENTS:** This handsome mushroom is both hallucinogenic and poisonous and should not be eaten. It obtained the common name, Fly Agaric because flies fed extracts from this mushroom became disoriented and died. The first time the authors collected this mushroom the caps were the size of dinner plates. Most species of *Amanita* are mycorrhizal.

## GRISETTE - *Amanita vaginata* group

**CAP:** Gray to tan, rounded becoming plane at maturity, edges striate, 6-9 cm (2 ½-3 ½ in). **GILLS:** White, crowded and free. **STALK:** Mostly white with tan/gray pattern, 6-10 cm (2 ½-4 in) X 1.0-1.5 cm (¼-½ in). Ring usually not present. Stalk emerges from a cup, which may be buried in the soil. **SPORES:** White, smooth, 8-10 µm, almost round. **HABITAT:** Collected in mixed coniferous/deciduous woods in the Black Hills. **EDIBILITY:** We recommend not eating any species of *Amanita* because of the many species in this genus that are deadly poisonous.
**COMMENTS:** The caps of most of our collections are a lighter gray/tan color than shown in the photograph above.

## DESTROYING ANGEL - *Amanita bisporigera* group

**CAP:** White, rounded when young becoming plane at maturity, sometimes with a slightly raised central portion, 5-10 cm (2-4 in). **GILLS:** White, crowded and free. **STALK:** White, 6-10 cm (2 ½-4 in) X 1-1.5 cm (¼-½ in); usually with a white, delicate skirt-like ring, which may disappear with age. Stalk emerges from a distinct cup, which may be partially buried in the soil. **SPORES:** White, smooth, 9-14 X 7-10 µm, almost round. **HABITAT:** Collected from soil in mixed conifer/deciduous areas where ironwood is present. It is reported as common throughout North America. **EDIBILITY: DEADLY POISONOUS!** It contains amanitin, a powerful toxin which destroys the liver and kidneys and can result in death. **COMMENTS:** This pure white mushroom is elegantly beautiful. Since the cup is one of the diagnostic characters, it is important to dig deep to retrieve it when collecting. Mushrooms with a ring, a cup, free gills and white spores are in the genus *Amanita*. Because many species of *Amanita* are very poisonous, **it is recommended that any mushroom with a ring, a cup, free gills, and white spores be avoided.**

## GROUP 1

### OYSTER MUSHROOM - *Pleurotus populinus*

**CAP:** White to tan, smooth, fan-shaped bracket, 7.5-15 cm (3-6 in). **GILLS:** White to cream, crowded and decurrent. **STALK:** Absent or if present, very short, eccentric or lateral. **SPORES:** White, smooth 7.5-12.3 X 2.5-4.3 μm. **HABITAT:** Grows shelf-like throughout the summer on standing or dead aspen in the Black Hills. **EDIBILITY:** Edible and very delicious and one of the authors' favorites. Best eaten when young, soft and fresh. **COMMENTS:** It acquired this common name because the fruiting bodies are shell-shaped and frequently have a fishy odor. Occasionally it will be found in large clusters. *Pleurotus ostreatus* is probably also present in the Black Hills and can only be separated from *P. populinus* by mating compatibility studies. Both are delicious.

GROUP 1

## FUNNEL CAP - *Clitocybe gibba*

**CAP:** Yellow to tan, smooth, deeply funneled, 6-8 cm (2-3 in). **GILLS:** White to cream, crowded and decurrent, tapering as they extend downward on the stalk. **STALK:** White to tan, 4-8 cm (1 ½-3 in) X 0.5-1.0 cm (¼-½ in). **SPORES:** White, smooth, 5-8 X 3-5 μm. **HABITAT:** Collected frequently throughout the summer in mixed coniferous/deciduous woods. **EDIBILITY:** Not considered poisonous, but not recommended because it can be confused with many other poisonous mushrooms. **COMMENTS:** It is one of the few mushrooms that clearly exhibits decurrent gills. The tan color, funneled cap and decurrent gills are good field characters to use.

# GROUP 1

## TINY TRUMPETS - *Xeromphalina campanella*

**CAP:** Yellow to gold to orange, funnel-shaped with inrolled margins, 0.5-2 cm (¼ -¾ in). **GILLS:** Yellow, distant, decurrent with veins connecting them. **STALK:** Yellow to brown, tough, 2-4 cm (¾-1 ½ in) X 3-5 mm (< ¼ in). **SPORES:** White, smooth, 5.5-7.0 X 3-4 μm. **HABITAT:** Grows in large groups (herds) on downed, dead conifer wood. **EDIBILITY:** Why bother? **COMMENTS:** The woody substrate, small size, decurrent gills, and bright color are distinguishing features. These mushrooms are very common, especially early in the summer and sometimes cover an entire well-rotted log. They can be collected throughout the summer. A similar species is *X. cauticinalis*, which grows on soil.

## False Chanterelle - *Hygrophoropsis aurantiaca*

**Cap:** Gold to orange, center depressed, margin may be inrolled, 5-8 cm (2-3 in). **Gills:** Same color as cap, thin, crowded, decurrent, usually dichotomously forked. **Stalk:** Shades of yellow, orange or tan, 5-8 cm (2-3 in) X 1.0 cm (½ in). **Spores:** White, smooth, 5-8 X 2.5-4.5 μm. **Habitat:** It is collected singly or in small groups on soil or decomposed wood in mixed coniferous/deciduous woods. **Edibility:** Not recommended. **Comments:** This beautiful golden mushroom is especially common after fall rains. It can easily be confused with the genus *Cantharellus*, the chanterelles, which are delicious and prized for the table. True chanterelles do not have typical thin gills, but have thicker structures that are defined as ridges. Chanterelle spores are 7-11 X 4-6 μm. *H. aurantiaca* can also be confused with the poisonous Jack-O-Lantern (*Omphalotus* species) mushroom. The Jack-O-Lantern Mushroom is more common in the Midwest and East, and does not have dichotomously forked gills and glows in the dark.

GROUP 1

## CONICAL WAXY CAP - *Hygrocybe conica*

This mushroom is similar to *Hygrophorous speciosus* (described on page **40**) with the following exceptions. The cap is more conical in shape, gills are attached to free, and the entire mushroom turns dark blue to black when bruised or with age. Spores can reach 14 μm in length. It is collected less frequently than *Hygrophorous speciosus*. **It is poisonous** and this is the reason it is recommended that no waxy caps should be eaten.

## GROUP 1

## PINE WAXY CAP - *Hygrophorous speciosus*

**CAP:** Round, wet to slimy, yellow/orange to red, 3-5 cm (1-2 in). **GILLS:** White, waxy, well-spaced, attached and usually weakly decurrent. **STALK:** Cream, 3-5 cm (1-2 in) X 0.5-1.0 cm (~ ¼ in). **SPORES:** White, smooth, 8-10 X 4.5-6 µm. **HABITAT:** Collected in groups on soil under conifers, especially pine. **EDIBILITY:** Reported as edible, but since it is small and because some other waxy caps are poisonous, it is not recommended. **COMMENTS:** These small mushrooms are common early in the season and observed less frequently throughout the summer. Many species of *Hygrophorous* are mycorrhizal.

## GROUP 1

## YELLOW WAXY CAP - *Hygrophorous chrysodon*

This mushroom is similar to *Hygrophorous speciosus* in shape, spore color, and edibility, except it is white to pale yellow with yellow flakes or granules on the cap and stalk, and is slightly larger. It is collected in mixed coniferous/deciduous habitats.

GROUP 1

## GRAY DUNCE CAPS - *Mycina pura*

**CAP:** Fragile, convex to conical shaped, lilac/gray to lilac/pink with striated edges, 2-4 cm (3/4-1 ½ in). **GILLS:** White to light gray, edges similar in color to surface of gills, well spaced. **STALK:** Fragile, gray, almost transparent, 3-4 cm (1-1 ½ in) X 2-4 mm (< ¼ in). **SPORES:** White, smooth, 7.5 X 3.5 µm. **HABITAT:** Our single collection was solitary on soil in a wet ravine under ironwood and aspen. **EDIBILITY:** Not recommended because of size and possibility of confusing it with other toxic species. **COMMENTS:** This species is a very beautiful, fragile mushroom. The cap is almost transparent, revealing the gills from the upper surface. Another species of *Mycena* that is similar and more commonly collected is *M. purpureofusca* and can be distiguished from *M. pura* because the edges of the gills are dark red to purple and it is more frequently collected on wood. Spores are 8-14 X 6-8.5 µm. Other species of *Mycina* can have tan-colored caps.

## GROUP 1

*Mycena* species

## HONEY MUSHROOM - *Armillaria ostoyae*

**GROUP 1**

**CAP:** Yellow to brown, sometimes with dark scales on upper surface, plane to slightly depressed at the center, 5-8 cm (2-3 in). **GILLS:** White to tan, attached to slightly decurrent. **STALK:** Shades of tan to brown with prominent white to yellow annulus, 5-8 cm (2-3 in) X 0.5-1.5 cm ($\frac{1}{4}$-$\frac{1}{2}$ in). **SPORES:** White, smooth, 6-10 X 5-6 μm. **HABITAT:** Frequently collected in mixed coniferous/deciduous woods and grows individually or in groups. **EDIBILITY:** Listed as edible and good, but use caution because there are some reports of gastronomical upsets. Be certain of your identification. **COMMENTS:** In the Black Hills, this mushroom most commonly fruits in the fall and can be collected in large numbers. It is a parasite primarily of pines and may be found growing in clusters at the base of a tree or growing as a saprobe from dead wood buried in the soil. It has also been called the Shoestring Fungus because it forms rhizomorphs in the soil or wood. Rhizomorphs are dark, tough, stringy mycelial modifications that resemble shoestrings. There is a great deal of variation in specimens seen in the Black Hills. *A. mellea* may also occur in this area.

GROUP 1

## BLEWIT - *Clitocybe nuda* (=*Lepista nuda*)

**CAP:** Pale pink/lavender/blue, edges rolled under when young, becoming plane or with a central knob when mature, 6-10 cm (2-4 in). **GILLS:** Same color as cap, crowded, attached to weakly decurrent. **STALK:** Same color as cap, 6-10 cm (2-4 in) X 2-3 cm (3/4-1 in). **SPORES:** Very light pink, slightly rough, 5.5-8 X 3.5-5 μm. **HABITAT:** Collected in moist wet canyons under mixed coniferous/deciduous woods. **EDIBILITY:** It is reported as edible, but is too beautiful to eat. **COMMENTS:** The Blewit is one of our most elegant robust mushrooms and collected infrequently in small groups. The delicate color can fade rather quickly after picking. There is another species, *Clitocybe glaucocana*, which resembles the Blewit, but is lighter colored and is smaller.

## GROUP 1

## WHITE LEUCOPAX - *Leucopaxillus albissimus*

**CAP:** White, dry, unpolished and suede-like, the edge inrolled when young, plane and centrally depressed when mature, 5-15 cm (2-6 in). **GILLS:** White to tan, attached to indistinctly decurrent. **STALK:** White, frequently becoming wider at the base, which typically has copious amounts of mycelium, 5-8 cm (2-3 in) X 1-2 cm ( ½-¾ in). **SPORES:** White, rough, 5.5-8.5 X 4-6 µm. **HABITAT:** It is found on soil singly or in groups in mixed coniferous/deciduous forests. **EDIBILITY:** Not recommended. Tough and bitter. **COMMENTS:** This mushroom is tough and robust. It is similar to some species of *Clitocybe* and *Tricholoma* and can only be distinguished using spore characters. Some species of *Melanoleuca* can be confused with the White Leucopax. Species of *Melanoleuca* have the same coloration, the suede-like cap surface, attached gills and white spores, but the stalk is evenly thick throughout, with no bulbous base or copious amount of mycelium. It is reported that some species of *Leucopaxillus* may be mycorrhizal.

## GROUP 1

### PLUMS AND CUSTARD - *Tricholomopsis rutilans*

**CAP:** Egg-yolk yellow streaked with reddish-purple fibrils, round to plane, 5-8 cm (2-3 in). **GILLS:** Yellow, attached. **STALK:** Cream to yellow, 5-8 cm (2-3 in) X 1-1.5 cm (½-¾ in). **SPORES:** White, smooth, 5-7 X 3-5 μm. **HABITAT:** On wood or rotten wood. **EDIBILITY:** Not recommended. **COMMENTS:** This is a very beautiful mushroom and when fibrils are distinct it is an easy one to identify. There are two other species that are similar. *T. decora* has a yellow cap with black to dark brown fibrils and typically grows in shelves with an eccentrically attached stalk. *T. platyphylla* has a white cap covered with grayish brown fibrils. It has white gills and a white stalk.

## GROUP 1

### Oak Gymnopus -
*Gymnopus dryophilus*
(=*Collybia dryophila*)
**CAP:** Shades of tan, plane becoming flattened with age, 3-5 cm (1-2 in) wide. **GILLS:** Attached, crowded and white. **STALK:** Tan-white, hollow and flattened near top, 5-7 cm (2-3 in) X 0.5-1 cm (~ ¼ in). **SPORES:** White, smooth, 5-7 X 2-3.5 µm. **HABITAT:** It is very commonly collected singly on soil in mixed conifer and aspen woods throughout the summer. **EDIBILITY:** Not recommended.
**COMMENTS:** This is one of the most common mushrooms that occurs throughout the Black Hills and during most of the summer. It is reported to occur in the east under oak, but we find it with pine and aspen. Another common mushroom collected throughout the summer and very similar to the Oak Collybia is *Gymnopus acervatus* (=*Collybia acervata*). This mushroom is found commonly in groups on wood. Occasionally *G. dryophilus* and *G. acervatus* show cream-colored

tumor-like growths on the fruiting bodies. They can occur on the cap, gills or stalk. The cause and effect of this malady are unknown.

## Brittle — White to Yellow-Spored Mushrooms

With some practice, the collector can recognize this group because of the stocky appearance, brittle texture and stiff gills that break easily. If this group of mushrooms is examined microscopically, the cap tissue contains sphaerocysts (spherical cells) intermingled with the filamentous cells.

### Burnt Sugar Mushroom - *Lactarius aquifluus*
**Cap:** Shades of orange to tan to brown with tiny fibrils, plane and slightly depressed in the center, 6-8 cm (2-3 in). **Gills:** Buff to light orange, brittle, attached and slightly decurrent, exuding a colorless latex. **Stalk:** Tan, hollow with orange fibrils, 5-7 cm (2-3 in) X 1.0 cm (½ in). **Spores:** White, rough with warts and ridges, 6-9 X 5.5-7.5 µm. **Habitat:** It is collected from *Sphagnum* bogs in the Hills. It is reported from bogs in northeastern North America. **Edibility:** We don't recommend it because *Paxillus involutus*, a similar but **poisonous** mushroom, grows in proximity with *L. aquifluus*

# GROUP 1

and at the same time. **COMMENTS:** The most distinguishing character of this mushroom is the strong and persistent maple syrup smell. Even after drying the smell is characteristic. It is important to collect and cut the gills of fresh specimens to observe the latex. Most species of *Lactarius* are mycorrhizal.

**GROUP 1**

## MILKY MUSHROOM - *Lactarius mammosus*
## (=*Lactarius hibbardae*)

**CAP:** The tan/gray cap is similar in size and texture to *Lactarius aquifluus*, but is slightly depressed in the center with inrolled edges. **GILLS:** Cream to buff to orange, brittle, attached to decurrent, exuding a white milky latex in fresh specimens. **STALK:** Tan to gray, 5-7 cm (2-3 in) X 1.0 cm (½ in). **SPORES:** White, rough with warts and ridges, 7.5-8.8 X 5.0-7.5 µm. **HABITAT:** This mushroom is collected from *Sphagnum* bogs in the Hills and has been reported from boggy areas in northeastern North America. **EDIBILITY:** Unknown. **COMMENTS:** The milky mushroom grows in the same areas as the Burnt Sugar Mushroom and has a pleasant, but not a maple syrup odor. The absence of the maple syrup smell and milky latex can separate these species. However, it should be noted that there are many other species of *Lactarius* with milky latex and some are poisonous. The specimen in the photo shows more gray than many collections.

# GROUP 1

## ROSY-RED RUSSULAS - *Russula* species

**CAP:** Variable shades of pink and red, may be sticky and wet when immature, round to plane, sometimes with a striated margin, 5-15 cm (2-6 in). **GILLS**: White, cream to yellow, crowded, brittle and attached. **STALK:** White to pink, 5-10 cm (2-4 in) X 1-3 cm (1/2- 1 in). **SPORES:** White, cream, pale yellow to yellow, with warts and ridges, 9-12 X 6.5-9 μm. **HABITAT:** This group of mushrooms is collected on soil in coniferous/deciduous woods. Many are associated with pine and spruce. **EDIBILITY:** Since several pink to red species of *Russula* are poisonous it is recommended that any pink to red *Russula* not be eaten. **COMMENTS:** We have grouped several similar species that have pink to red caps into this group. *Russula* species are extremely variable in cap color, which may change during drying. Members of this group are frequently encountered in the Hills late in the summer and are some of our most beautiful mushrooms. The *Russula emetica* group is a part of the Rosy-Red Russula group. These common, poisonous mushrooms have a red cap with striated edges, white gills, white spores, and a white stalk. Most species of *Russula* are mycorrhizal.

## GROUP 1

### VARIABLE RUSSULA - *Russula alutacea* group

**CAP:** Yellow to orange to tan to red, round to plane, sometimes centrally depressed, smooth, wet and sticky when immature, striated around the edges, the top layer very thin and tearing around the edges, 6-12 cm (2-5 in). **GILLS**: Cream to yellow, brittle, almost distant and attached. **STALK:** White, smooth, 5-8 cm (2-3 in) X 1-2 cm (½-¾ in). **SPORES:** Yellow with warts, 8-11 X 6.5-9 μm. **HABITAT:** It fruits on soil in mixed coniferous/deciduous woods. **EDIBILITY:** It is not recommended because many species may be included in this group and could be confused with poisonous species. **COMMENTS:** It occurs late summer and is difficult to identify without using chemicals such as phenol which stains the mushroom purple-red. The colors are extremely variable, showing yellows, oranges, tans and reds. The authors have observed squirrels eating and stockpiling this group of mushrooms as well as mushrooms in the Rosy Russula group.

**GROUP 1**

## FRAGILE RUSSULA - *Russula fragilis*

**CAP:** White with shades of pink, wet and sticky, upper layer very thin exposing a view of the gills, plane, margins uplifted, 4-6 cm (1 ½-2 in). **GILLS:** White, brittle, attached. **STALK:** White, smooth, 5-7 cm (2-3 in) X 1 cm (½ in). **SPORES:** White, warts and ridges, 6-9 X 5-8 µm. **EDIBILITY:** Not recommended. **HABITAT:** Collected from soil in mixed coniferous/deciduous woods. **COMMENTS:** This is a delicate, beautiful mushroom, smaller than most other *Russula* species.

GROUP 1

## BLACKENING RUSSULA - *Russula albonigra*

**CAP:** Hard, tough, white when immature, turning gray and finally black at maturity, depressed in center, margin inrolled, 5-8 cm (2-3 in). **GILLS:** White when immature, finally changing to black when mature, attached and brittle. **STALK:** White when immature, finally becoming black, 4-7 cm (1 ½-2 ½ in) X 1-2 cm (½-¾ in). **SPORES:** White, with warts and ridges, 7-10 X 5.5-7.5 µm. **HABITAT:** It is found on soil in mixed coniferous/deciduous woods and is uncommon. **EDIBILITY:** Not recommended because easily confused with poisonous species. **COMMENTS:** The dramatic change from white to black is the distinguishing feature of this mushroom and until this is observed, it can be confused with other species of *Russula*. This change occurs over several hours. The first collection observed was white when collected and totally black the next day.

## SHORT-STEMMED RUSSULA - *Russula brevipes*

**CAP:** White to tan with soil particles attached, very tough, deeply depressed in the center, 8-10 cm (3-4 in). **GILLS:** Upswept appearance, white to tan, brittle, attached to decurrent. **STALK:** White and tough, 2-5 cm ($3/4$-2 in) tall X 2-4 cm ($3/4$-1 $1/2$ in) wide. **SPORES:** White, with warts and ridges, 8-11 X 6.5-9 μm. **EDIBILITY:** Too tough to be appealing. **HABITAT:** *Russula brevipes* is common and collected on soil from mixed coniferous/deciduous woods. **COMMENTS:** This robust, tough mushroom with brittle gills has a very short stipe, which is almost as wide as tall. Many times only the surface of the cap is visible, the remainder buried in the soil.

GROUP 1

# PINK-SPORED MUSHROOMS

## FAWN MUSHROOM - *Pluteus cervinus*

**CAP:** White to tan, smooth or with discrete radial fibrils, plane, 5-6 cm (2-2 ½ in). **GILLS:** Pink, crowded and free, sometimes exhibiting an uplifted appearance. **STALK:** White to tan, 5-7 cm (2-3 in) X 0.5-1 cm (¼-½ in). **SPORES:** Salmon to pink, smooth, almost round, 5.5-7 X 4-6 μm. On gill edges are microscopic pear-shaped cells called cystidia and they display long necks with horns at the tapered end. **HABITAT:** Occurs singly on dead wood. **EDIBILITY:** Listed as edible, but loses desirable, firm texture quickly. **COMMENTS:** This is an elegant mushroom. The pink, free gills are unusual. It can be collected early in the spring and throughout the summer, but never in large quantities. Another species of *Pluteus* collected frequently is *P. admirabilis*. This smaller mushroom, also collected on wood has a yellow cap and stalk, but also exhibits the pink, free gills and salmon to pink spores.

## NETTED MUSHROOM - *Rhodotus palmatus*

**CAP:** Salmon to pink, round, netted, margin inrolled, usually exhibiting a bracket-like growth habit, 5-8 cm (2-3 in). **GILLS:** Salmon to pink, attached. **STALK:** White to pink, eccentric or lateral, 5-8 cm (2-3 in) X 0.7 cm ($1/4$ in). **SPORES:** Pink, warted, 6-8 µm, almost round. **HABITAT:** It is reported to grow only on hardwoods and was collected in this study primarily on ironwood and boxelder, but also on pine. **EDIBILITY:** It is unknown, but it is too beautiful to eat. **COMMENTS:** This is a beautiful and unusual mushroom. It is reported as uncommon in North America and is protected in parts of Europe. Heavy

**GROUP 1**

fines are assessed if picked in some countries. This clearly shows the need to not remove downed wood because it can be an important substrate for rare fungi.

## BLACK-SPORED MUSHROOMS

In addition to black spores, mushrooms featured in this group have round to sausage-shaped cells in the cuticle (outside layer) of the cap (pileus).

### BELL-SHAPED PANAEOLUS - *Panaeolus campanulatus* group

**CAP:** Smooth, shades of brown, bell-shaped to conical, sometimes with white veil fragments attached to margin, 2-3 cm ($^3/_4$-1 in). **GILLS:** Mottled gray to black, becoming black at maturity, attached. **STALK:** Tan and brittle, 5-7 cm (2-2 $^3/_4$ in) X 3-5 mm (<$^1/_4$ in). **SPORES:** Black, smooth, 13-18 X 7.5-12 μm. **HABITAT:** It is found on dung or in grass where livestock have grazed. **EDIBILITY:** Not recommended. **COMMENTS:** It can be confused with some very poisonous little brown mushrooms.

# GROUP 1

The cap is very soft and can be broken into little pieces easily. The group photo is of immature specimens. Mature specimens appear more like the mushroom on the left.

### SHAGGY MANE - *Coprinus comatus*
**CAP:** White-gray, scaly, cylindrical, 2-5 cm (1-2 in) wide X 5-10 cm (2-4 in) tall. **GILLS:** Almost free, gray turning black and deliquescing (self-digesting) at maturity. **STALK:** White to gray, sometimes exhibiting a ring, 5-15 cm (2-6 in) X 0.5-2 cm (1/4-3/4 in). **SPORES:** black, smooth, 10-16 X 7-9 µm. **HABITAT:** It occurs on soil in grassy areas, woods, along roadsides and in a variety of habitats. **EDIBILITY:** Edible, but only when young,

GROUP 1

firm and white prior to spore production. **COMMENTS:** This mushroom is also called "Lawyer's Wig". The scaly cap gives it the shaggy appearance. Shaggy Manes emerge, mature and deliquesce quickly.


**MICA CAP** - *Coprinus micaceus*
**CAP:** Tan to brown, exhibiting glistening particles when young, bell-shaped, 2-4 cm (¾-1 ½ in) wide X 4-5 cm (1 ½-2 in ) tall.
**GILLS:** Almost free, gray turning black and deliquescing to an inky ooze at maturity. **STALK:** White to gray, 6-10 cm

61

## GROUP 1

(2 ½-4 in) X ~ 5 mm (~ ¼ in). **SPORES:** Black, smooth, 7-11 X 4-6 μm. **HABITAT:** Common in a variety of habitats to include grass, woody debris or buried wood. **EDIBILITY:** Not recommended. **COMMENTS:** It is common to see this fungus growing in groups in grassy areas and in areas where trees have been cut down. The group emerges and deliqueses quickly forming inky masses. There is another very similar and common species that deliqueses called the **Common Inky Cap**, *Coprinus atramentarius*. This mushroom does not have glistening particles when young and has a thicker stalk. The **Common Inky Cap** (also known as

Tippler's Bane) is reported as edible when young, firm and white before spores form **but should not to be eaten with any form of alcohol because the symptoms caused by mixing alcohol and a chemical in these mushrooms are similar to those experienced by individuals consuming alcohol after treatment with disulfiram (Antabuse)**. Still another less common, but similar mushroom that deliqueses is the elegant **Scaly Inky Cap**, *C. quadrifidus*. The gray to tan striated cap

*C. quadrifidus*

*C. niveus*

*C. lagopus*

has distinct buff flakes. Another similar deliquesing mushroom is the smaller **Wooly Inky Cap**, *C. lagopus*. Conical to plane caps are striated. Both the cap and stalk are white and covered with delicate white to gray hairs. Spores are 10-16 X 6-7 μm and are smooth. *C. niveus* is a species that occurs on dung. It too is white to gray with a powdery or mealy cap.

## GROUP 1

# PURPLE BROWN-SPORED MUSHROOMS

### WOODLAND AGARICUS - *Agaricus silvicola*

**CAP:** White to tan, smooth, sometimes with scales, round to plane at maturity, staining light yellow when bruised, 8-10 cm (3-4 in). **GILLS:** Light colored becoming dark brown in age, crowded and free. **STALK:** White to tan, smooth, 7-9 cm

(3-3 ½ in) X 1.0-1.5 cm (½-¾ in), usually with a white skirt-like ring, which may be brown if covered with spores. **SPORES:** Chocolate to purple brown, smooth, 5-6.5 X 3.5-4 µm. **HABITAT:** This mushroom occurs solitary or in small groups on soil in mixed coniferous/deciduous woods in the Hills and is reported as widely distributed and common in North America. **EDIBILITY:** Not recommended. **COMMENTS:** It can be easily confused with the Meadow Mushroom, *Agaricus campestris* (spores smooth, 6.5-8.5 X 4-5.5 µm) which is edible and very good. The woodland habitat of *A. silvicola* is one character to use when determining its identity.

Another interesting species of *Agaricus* that is very good to eat is *A. bitorquis* (spores smooth, 5-7 X 4-5.5 µm). It is unusual because it is found along roadsides among weeds and where the ground is very disturbed. One feature of this mushroom

*A. bitorquis*

is that it never entirely emerges above the soil. To collect them you have to dig them out. If you are interested in eating them be sure to collect young specimens with pink gills and check for millipedes and insects that also like to eat them.

# SCALY STROPHARIA - *Stropharia kauffmanii*

**CAP:** Dry, yellow to tan, with yellowish-brown scales, round to plane at maturity, 6-10 cm (2-4 in). **GILLS:** Purple brown to purple gray, edges often eroded, attached. **STALK:** White to tan, scaly, with a prominent white ring which may be darkened with the spores, 6-9 cm (2- 3 $\frac{1}{2}$ in) X 1-2 cm ($\frac{1}{2}$-$\frac{3}{4}$ in). **SPORES:**

Purple brown, smooth, 6-8 X 4-4.5 μm. **HABITAT:** Grows singly on soil near decayed wood in mixed coniferous and deciduous woods. **EDIBILITY:** Not recommended. **COMMENTS:** A very beautiful mushroom, appearing early in the season. Sometimes partial veil remnants will adhere to the edge of the cap giving it a frilly appearance.

## COMMON WOODLAND PSATHYRELLA - *Psathyrella candolleana*

**CAP:** White to gray with gray or tan fibrils, round to plane at maturity, 5-7 cm (2-3 in). **GILLS:** White becoming purplish gray to brown to dark brown at maturity, attached. **STALK:** White and scurfy, sometimes with remnants of a ring, 6-8 cm (2-3 in) X 0.5-1 cm (¼ -½ in). **SPORES:** Purple brown, smooth, 7-10 X 4-5 μm. **HABITAT:** It is found scattered on soil in mixed coniferous/deciduous woods, and is also reported from lawns. **EDIBILITY:** Not recommended because it can be confused with other dark-spored inedible mushrooms. **COMMENTS:** It appears early in the season and is fairly common throughout

GROUP 1

the summer. The top of the cap is soft and can be readily torn into little pieces. The outer layer of the cap is composed of round cells. The single specimen shown here is immature.

**GRACEFUL PSATHYRELLA** - *Psathyrella gracilis*
**CAP:** Conical to bell-shaped, brown, 2-3 cm (³⁄₄-1 in). **GILLS:** Gray, becoming dark brown at maturity, attached. **STALK:** White, smooth, thin, 6-8 cm (2-3 in) X 3-5 mm (¼ in). **SPORES:** Purple brown, smooth, 10-15 X 6-8 μm. **HABITAT:** Found in groups on or near dead wood, wood chips, or decomposed

## GROUP 1

wood. **EDIBILITY:** Not recommended. **COMMENTS:** This fragile mushroom has a soft, thin cap, with a cuticle composed of round cells and the gills can be observed through the cap.

GROUP 1

# Orange Brown, Rusty Brown to Brown-Spored Mushrooms

## Common or Boring Gymnopilus - *Gymnopilus sapineus*

**CAP:** Golden yellow to orange brown at maturity, convex to plane, dry, 2-4 cm (¾-1 ½ in). **GILLS:** Rusty yellow to rusty brown when mature, attached. **STALK:** Tan with orange fibrils, 3-6 cm (1-2 ¼ in) X 0.5-1.0 cm (¼-½ in). **SPORES:** Orange to rusty brown, almost rough, 7-10 X 4-5.5 µm. **HABITAT:** Occurs singly or in groups on wood. **EDIBILITY:** Not recommended. **COMMENTS:** This is a common mushroom and is distinguished by the orange color, orange spores, dry cap and growth on wood. It can be confused with some species of *Cortinarius*, but *Cortinarius* species occur on soil.

GROUP 1

## VERNAL CHANGING PHOLIOTA - *Pholiota vernalis* (=*Kuehneromyces vernalis*)

**CAP:** Yellow to brown, hygrophanous (color changing as it dries), typically with the periphery drying darker brown, smooth, bell-shaped to conical, sometimes with a central knob, becoming flat at maturity, 2-3 cm ($3/4$-1 in). **GILLS:** Yellow to brown, attached. **STALK:** Tan to brown, with indistinct fibrils, 4-7 cm (1 $1/2$-2 $3/4$ in) X 3 mm (<$1/4$ in). **SPORES:** Brown, smooth,

5.5-7.5 X 3.5-5 μm. **HABITAT:** These mushrooms grow in groups on dead, rotten wood. **EDIBILITY:** Not recommended. It can be confused with some deadly poisonous little brown mushrooms. **COMMENTS:** It usually is collected in early spring in cool weather.

## SILKY LITTLE BROWN MUSHROOM - *Inocybe fastigiata*

**CAP:** Shades of tan to brown, conical to bell-shaped when young, expanding and becoming plane as it matures, but retaining a distinct central knob (umbo), silky texture, margin splitting, 3-5 cm (1-2 in). **GILLS:** Attached, tan to brown at maturity with yellow tints. **STALK:** Cream to tan, 4-6 cm (1 ½-2 ½ in) X 5-7 mm (~ ¼ in). **SPORES:** Dull brown, 10-13 X 5-8 μm. **HABITAT:** This mushroom occurs on the soil singly or in small groups in mixed coniferous/deciduous forests. **EDIBILITY:** It is listed as poisonous and not recommended for eating. No species of *Inocybe* should be eaten as some species contain muscarine. **COMMENTS:** This little mushroom is one of the most commonly collected species and occurs throughout the season. The knob, dull brown spores and silky texture of the cap are

good characters for identification. Fresh specimens smell like mild green corn. A similar species, *I. geophylla* also occurs in the Black Hills and is almost white. Most species of *Inocybe* are mycorrhizal.

## BOGGY CORTINARIUS - *Cortinarius* species

**CAP:** Reddish brown, wet and sticky, round to plane at maturity, sometimes with a central knob, immature with white silky fibrils, 4-7 cm (1 ½-3 in). **GILLS:** Reddish brown at maturity, attached, distant. **STALK:** White to tan with rusty brown fibrils, club-shaped, 4-8 cm (1½-3 in) X 1- 1.5 cm (½ -¾ in), when immature exhibiting a white cobwebby veil, usually leaving a white fibrous ring. **SPORES:** Rusty to cinnamon brown, minutely rough, 7-11 X 4-6 μm. **HABITAT:** It grows on soil in mossy areas with bog birch and spruce. **EDIBILITY:** Not recommended because many species are poisonous. **COMMENTS:** One of many species of *Cortinarius* that occur in a variety of habitats. Most species of *Cortinarius* are mycorrhizal.

## SLIMY CORTINARIUS - *Cortinarius* species

**CAP:** Orange to tan, sticky and slimy, flat to round with a central knob, 4-6 cm (1 ½-2 ½ in). **GILLS:** Tan to gray, becoming dark brown at maturity, attached. **STALK:** White to tan, 5-9 cm (2-3 ½ in) X 1-1.5 cm (½-¾ in), displaying a cobwebby or silky

# GROUP 1

veil when immature and remnants of the veil may remain as orange to brown fibrils. **SPORES:** Rusty brown, rough, 11-12 X 5-6.5 μm. **HABITAT:** It occurs singly on soil in mixed coniferous/deciduous woods. **EDIBILITY:** Not recommended because many *Cortinarius* species are deadly poisonous. **COMMENTS:** There are numerous species in the genus *Cortinarius*, and most are difficult to identify and require microscopic characters. There are several species with slimy caps. A mushroom with rusty brown spores, occurring on soil and displaying a cobbwebby veil usually will be a *Cortinarius* species. Some species have lavender or purple colors blended with tan or orange.

# GROUP 2 - *Key to Boletes in Field Guide*

**1A.** Cap distinctly yellow, usually wet and sticky; pore surface distinctly yellow ................................................................
................Granulated Slippery Jack *(Suillus granulatus)* p. 77
**1B.** Cap and pore surface not yellow ................................................ 2
**2A.** Cap brown, dry, sometimes cracking and exposing white flesh; pore surface cream to tan to gray ..............
................................................Brown Bolete *(Leccinum snellii)* p. 78
**2B.** Cap with shades of orange to red/brown; pore surface white to yellow ........................................................................... 3
**3A.** Cap burnt orange; stalk white with numerous tiny orange to black scabers; under aspen ................................................
................................................Aspen Bolete *(Leccinum insigne)* p. 79
**3B.** Cap orange brown to red brown; stalk white with fibrils the same color as the cap; base usually bulbous ................
................................................King Bolete *(Boletus edulis)* p. 80

# GROUP 2 - *Boletes*

## GRANULATED SLIPPERY JACK - *Suillus granulatus*

**CAP:** Yellow to yellow-brown, round to plane, soft, wet and sticky, 5-8 cm (2-3 in). **PORES:** Underneath cap are yellow pores. **STALK:** White to yellow with small reddish to brown glandular dots, 4-7 cm (1 ½-2 ¾ in) X 1-1.5 cm (½-¾ in). **SPORES:** Brown to ochre, smooth and spindle-shaped, 7-10 X 2.5-4 µm. **HABITAT:** It grows in soil in coniferous woods, especially with pine. **EDIBILITY:** Listed as edible, but texture is unappetizing so we do not recommend it. **COMMENTS:** This is the most common bolete collected in the Hills, which appears early summer and can be collected throughout the growing season. Mature, dry specimens may not have sticky caps, but debris remains on the cap, indicating it was sticky. Species of *Suillus* are mycorrhizal.

GROUP 2

## BROWN BOLETE - *Leccinum snellii*

**CAP:** Dry, brown, round, sometimes with a cracked cap exposing white flesh, 5-7 cm (2-3 in). **PORES:** Cream to tan to gray, almost free from stalk. **STALK:** Cream to tan covered with dark brown to black scabers (tufted hairs), 5-7 cm (2-3 in) X 1.0 cm (½ in). **SPORES:** Brown, spindle-shaped, 14-20 X 5-7 μm. **HABITAT:** Occurs singly on soil in mixed aspen/coniferous woods. **EDIBILITY:** Unknown. **COMMENTS:** This is an elegant uncommon mushroom collected mid to late summer. Caps of our collections are cracked appearing like cookies cracking during baking. This is probably due to dry conditions. Species of *Leccinum* are listed as mycorrhizal.

## GROUP 2

## ASPEN BOLETE - *Leccinum insigne*

**CAP:** Burnt orange, round to plane at maturity, 5-7 cm (2-2 ¾ in). Periphery may have white remnants of tissue overhanging the margin. **PORES:** White, with myriads of tiny pores almost free of stalk. **STALK:** White, with numerous orange/brown/black scabers, 8-10 cm (3-4 in) X 2-4 cm (¾-1½ in). **SPORES:** Yellow to brown, smooth, spindle-shaped, 11-18 X 4-6 μm.
**HABITAT:** Collected singly on soil, most commonly under aspen. **EDIBILITY:** It is listed as edible. **Be careful as there are poisonous look alikes! COMMENTS:** This is a handsome bolete and one of the few commonly collected. The flesh stains gray to black when bruised.

## KING BOLETE - *Boletus edulis*

**CAP:** Burnt orange to red/brown, smooth, round to plane at maturity, 10-15 cm (4-6 in). **PORES:** White to yellow at maturity. **STALK:** White with fine veins the same color as the cap, usually with a bulbous base, 7-10 cm (3-4 in) X 5-7 cm (2-3 in). **SPORES:** Brown, smooth, spindle-shaped, 13-19 X 4-7 µm. **HABITAT:** It is collected singly on soil in mixed deciduous/coniferous woods. **EDIBILITY:** Listed as edible and reported as one of the best! **COMMENTS:** This is a rather large, stocky bolete and very beautiful. It is collected infrequently during July and August. It is called porcini or ceps in European cookbooks. Most species of *Boletus* are mycorrhizal.

# GROUP 3 - Key to Polypores, Toothed Fungi, Jelly Fungi, Ridged Fungi, Resupinate Fungi, Coral and Club Fungi in Field Guide

**1A.** Fruiting body, usually tough and woody, growing as a bracket or shelf, attached to substrate laterally ... 2

**1B.** Fruiting body tough, growing upright and attached centrally to eccentrically, or flat, or entire fruiting body elastic and "jelly-like" ... 10

**2A.** Lower surface consisting of elongated pores resembling gills of a mushroom ... Rusty-Gilled Polypore *(Gloeophyllum sepiarium)* p. 84

**2B.** Lower surface consisting of round to hexagonal pores or lower surface covered by a membrane hiding the pore surface ... 3

**3A.** Lower surface covered by a membrane hiding the pore surface ... Cryptic Globe Fungus *(Cryptoporus volvatus)* p. 85

**3B.** Lower pore surface exposed, consisting of round to hexagonal pores ... 4

**4A.** Fruiting body perennial, thick and consisting of more than one layer of pores (cut longitudinally to determine) ... 5

**4B.** Fruiting body annual, thinner with only one layer of pores ... 6

**5A.** Fruiting body lower surface white, consisting of very small pores; upper surface with brownish red zone ... Red-Belted Conk *(Fomitopsis pinicola)* p. 86

**5B.** Fruiting body hoof-shaped; lower surface brown, upper surface gray/brown to black ... False Tinder Polypore *(Phellinus tremulae)* p. 87

**6A.** Upper and lower surfaces bright orange-red ... Red Polypore *(Pycnoporus cinnabarinus)* p. 88

**6B.** Upper surface rusty brown, brown, tan, cream or gray ... 7

**7A.** Upper surface rusty brown, hairy; lower surface tan to yellowish brown to dark brown; sometimes growing flat ... Rusty Polypore *(Inonotus rheades)* p. 89

**7B.** Upper surface brown, tan, cream or gray ... 8

**8A.** Upper surface gray, lower surface smoky gray consisting of very tiny pores ............ Smoky Polypore *(Bjerkandera adusta)* p. 90

**8B.** Upper surface brown, tan, cream; lower surface white, tan, yellow or brown ............ 9

**9A.** Upper surface of the fan-shaped bracket with distinctly colored zones ............ Turkey Tails *(Trametes versicolor)* p. 91

**9B.** Upper surface not distinctly zoned, hairy, cream, or tan ............ Hairy Turkey Tail *(Trametes hirsuta)* p. 92

**10A.** Lower surface consisting of pores; stalk present or not ... 11

**10B.** Lower surface toothed, ridged, smooth (branched or unbranched), wrinkled, or entire fruiting body jelly-like ............ 14

**11A.** Occurring singly, small, 2 in. or less, distinct stalk attached centrally ............ 12

**11B.** Fruiting body larger; stalk indistinct or not evident, attached centrally to eccentrically ............ 13

**12A.** Upper surface with concentric zones, centrally depressed, periphery uneven ............ Fairy Stool *(Coltricia perennis)* p. 93

**12B.** Upper surface not zoned, not centrally depressed, periphery hairy ............ Fringed Polypore *(Polyporus arcularius)* p. 94

**13A.** Upper surface mottled, hairy; lower surface shades of green to tan; stalk not evident, attached centrally ............ Dye Polypore *(Phaeolus schweinitzii)* p. 95

**13B.** Lower surface white to tan; upper surface brown and almost black in the center; base of indistinct stalk black, usually attached eccentrically ............ Black-Leg *(Polyporus badius)* p. 96

**14A.** Lower surface toothed ............ Scaly-Toothed Fungus *(Sarcodon imbricatum)* p. 97

**14B.** Lower surface ridged, smooth (branched or unbranched); wrinkled or entire fruiting body jelly-like ............ 15

**15A.** Fruiting body elastic and jelly-like ............ 16

**15B.** Fruiting body ridged, smooth (branched or unbranched), or wrinkled ............ 17

**16A.** Yellow to orange colored ............ Witch's Butter *(Tremella mesenterica)* p. 98

**16B.** Brown colored, ear-shaped
   Brown Squirrel Ears *(Auricularia auricula)* p. 99
**17A.** Lower surface ridged
   Pig's Ears *(Gomphus clavatus)* p. 100
**17B.** Lower surface smooth (branched or unbranched) or wrinkled ... 18
**18A.** Lower surface smooth; fruiting body highly branched or club-shaped ... 19
**18B.** Entire fruiting body flat and wrinkled, pink/orange
   Warts on Wood *(Peniophora rufa)* p. 101
**19A.** Highly branched
   Common Coral *(Ramaria apiculata)* p. 102
**19B.** Club shaped
   Truncate Club Coral *(Clavariadelphus truncatus)* p. 103

# GROUP 3 - *Polypores*

## RUSTY-GILLED POLYPORE - *Gloeophyllum sepiarium*

**FRUITING BODY:** A stalkless, leathery, bean-shaped bracket, 5-8 cm (2-3 in), upper surface rusty to dark brown, lower surface yellow to brown with pores modified to resemble thick plates or gills. **HABITAT:** It is observed on dead conifers. **EDIBILITY:** Too tough to consider. **COMMENTS:** This is a decomposer, one of our most common polypores, and causes a brown cubical rot on conifer and sometimes on aspen. Since it is an annual polypore it produces only one layer of pores.

## CRYPTIC GLOBE FUNGUS - *Cryptoporus volvatus*

**FRUITING BODY:** An annual, stalkless, hoof-shaped to spherical bracket, 2 -5 cm (¾-2 in) wide X 2-4 cm (¾-1 ½ in) tall; upper surface orange to brown, fading to cream with age, and cracking when dried; lower surface covered with a tough tan membrane, hiding the inner pore layer. **HABITAT:** Found on dead, standing pines. **EDIBILITY:** Too tough to consider. **COMMENTS:** This small fungus is quite unique because the pore layer is covered with a tough membrane, which if removed shows a salmon-pink layer of pores which form salmon to pink spores. Spores are deposited on the inside of the lower membrane. Insects are responsible for boring holes in the membrane to release the spores. It is easy to confuse this fungus with a very young Red-belted Conk (*Fomitopsis pinicola*), but *C. volvatus* is not as tough and woody and has the covered pore surface.

## RED-BELTED CONK - *Fomitopsis pinicola*

**FRUITING BODY:** A tough, woody stalkless bracket with incurved margins, 8-25 cm (3-10 in) X 4-8 cm (1½-3 in) thick; upper surface red to brown, sometimes with a red belt around the periphery; lower surface is white to tan and consists of myriads of tiny pores where spores are produced. **HABITAT:** Grows on dead and sometimes living conifers. **EDIBILITY:** Too tough to consider. **COMMENTS:** This is the most common bracket or conk in the Black Hills and can be seen anytime of the year. Nearly anywhere pine or spruce is growing, you will observe the Red-Belted Conk. This fungus is described as perennial because a new layer of pores is formed each year. If the fruiting body is cut longitudinally you can count the layers of pores and age the conk. It is a decomposer and associated with brown cubical sap rot, developing cracks in decayed wood. A much less common species is *F. cajanderi*, the Rosy Conk which has a pink pore layer.

GROUP 3

## FALSE TINDER POLYPORE - *Phellinus tremulae*

**FRUITING BODY:** Hard, perennial, hoof-shaped bracket, 10-13 cm (4-5 in) wide X 8-12 cm (3-4 ½ in) tall; upper surface dark gray or brown becoming black and cracked with age. Lower pore surface brown. If cut in longitudinal section the pores can be observed to be stuffed with white mycelium. You may need a hand lens to see this feature. **HABITAT:** Collected on aspen. **EDIBILITY:** Too tough to consider. **COMMENTS:** It causes a white trunk rot of living aspen and decays dead trees. The true Tinder Polypore, *Fomes fomentarius* is collected frequently on deciduous wood.

## RED POLYPORE - *Pycnoporus cinnabarinus*

**FRUITING BODY:** Tough, woody, round to fan-shaped bracket, 5-7 cm (2-3 in) X 1 cm (< ½ in) thick; upper surface orange to red; orange to red lower surface consists of many small pores. **SPORES:** White. **HABITAT:** It occurs singly or in small groups on dead wood, especially bur oak. **EDIBILITY:** Too tough to be edible. **COMMENTS:** This is a very easy polypore to recognize because it is bright red and has a single layer of pores. It causes a white rot of hardwoods.

## GROUP 3

### RUSTY POLYPORE - *Inonotus rheades*

**FRUITING BODY:** Hard, woody bracket (sometimes resupinate), exposed surface rusty brown, densely hairy, 2.0-10 cm (¾-4 in) in diameter; lower surface 1½-2 cm (½-¾ in) thick, tan to yellowish brown to dark brown consisting of many small pores. **HABITAT:** It is found most commonly in groups on dead aspen. **EDIBILITY:** Too tough and woody to consider. **COMMENTS:** This fungus is an annual although it appears thicker than most annuals. It causes a white heart rot of aspen.

## SMOKY POLYPORE - *Bjerkandera adusta*

**FRUITING BODY:** Thin, overlapping clusters of brackets, 5-7 cm (2-2 ½ in); upper surface white to tan to grays or browns; lower surface composed of very fine pores, distinctly smoky gray. **HABITAT:** This fungus grows in shelving groups on primarily dead deciduous wood, especially aspen. **EDIBILITY:** Too tough and woody. **COMMENTS:** The smoky gray pore surface is a distinguishing feature. It has a single layer of pores. It causes a white rot of hardwoods and sometimes conifers.

## GROUP 3

## TURKEY TAILS - *Trametes versicolor*

**FRUITING BODY:** Thin, brown to gray fan-shaped brackets, 5-7 cm (2-3 in) wide; upper surface with distinctly colored concentric zones, zones sometimes edged in white; lower surface consisting of fine white to tan pores. **HABITAT:** This handsome bracket grows in vertical and horizontal groups on primarily dead deciduous wood. **EDIBILITY:** Too tough and woody to be of value. **COMMENTS:** The multicolored zones and fan-shaped fruiting body gives this fungus its common name. It is handsome and distinctive. It is less common than the Hairy Turkey Tail, exhibits more distinct differences of coloration in concentric zones and is much less hairy. It causes a white rot of dead hardwoods.

GROUP 3

# Hairy Turkey Tail - *Trametes hirsuta*

**Fruiting Body:** Thin, overlapping clusters of brackets, 5-7 cm (2-3 in); attached laterally, in shades of white to yellow to tan; upper surface distinctly hairy, usually weakly concentrically zoned; lower surface with many small pores in shades of white, tan, yellow to brown. **Habitat:** It is found usually in groups on dead wood. **Edibility:** Too tough and woody. **Comments:** A common bracket collected on both deciduous and coniferous wood. It decays the sapwood of dead hardwoods.

24/2017 4:16 PM    Sales Receipt #64749
tore: WIC    Workstation: 1

# Wind Cave National Park
26611 US Hwy 385
Hot Springs SD 57747
Black Hills Parks & Forests Association
www.blackhillsparks.org
605-745-1263

ashier: Larkin

| em # | Qty | Price | Ext Price |
|---|---|---|---|
| 64 | 1 | $18.95 | $18.95 T |

MUSHROOMS AND

|  | Subtotal: | $18.95 |
|---|---|---|
| Non City Group | 6 % Tax: | + $1.14 |
|  | **RECEIPT TOTAL:** | **$20.09** |

mount Tendered: $20.10
Change Given: $0.01

Cash: $20.10


Become a member and save up to 15% on
purchases at cooperating public land agencies!
Only $30.00 - makes a great gift!

not satified, products may be returned to original
sales outlet within 30 days of purchase
oduct exchange or store credit will be issued after
30 days
All items must be in the original condition and
ackaging. All items must be accompanied by the
receipt.

Celebrating the 100th Anniversary of the National
Park Service!
Find Your Park

Thank you for supporting public lands!
Black Hills Parks & Forests Association
Since 1946

64749

## Wind Cave National Park
 26611 US Hwy 385
 Hot Springs SD 57747
 Black Hills Parks & Forests Association
 www.blackhillsparks.org
 605-745-7020

| Item | Qty | Price | Ext Price |
|---|---|---|---|
| | 1 | $19.95 | $19.95 T |

SD USE TAX AMT

| Subtotal | $19.95 |
| Sls Tax Amt | $2.13 |
| **RECEIPT TOTAL** | **$22.08** |

Amount Tendered: $22.08
Change Given: $0.00

Visa: $22.08

Become a member and save up to 15% on
all merchandise. Ask about our membership today.

Your admission tags may be exchanged for a refund,
store credit, exchange or merchandise. Please allow
up to 30 days for this process.

All items must be in the original condition with
packaging. All refunds must be accompanied by the
receipt.

Thank you for supporting the educational efforts of the Black Hills Parks & Forests Association.
Staff ID:

GROUP 3

## FAIRY STOOL - *Coltrichia perennis*

**FRUITING BODY:** Small, thin, dry polypore with a circular cap, 3-5 cm (1-2 in); uneven margin, central portion depressed; upper surface rust-brown with concentric rings; buff to gray lower surface consisting of decurrent pores. **STIPE:** Central and brown, 1-1½ cm (~½ in) tall X 4-7 mm (~¼ in) wide. **HABITAT:** One of our collections was found on soil along an old logging road in coniferous woods. **EDIBILITY:** Too tough to consider. **COMMENTS:** When fresh, it is reported that the cap is velvety. The fungus is reported as widespread, along paths and road banks, but uncommon.

GROUP 3

# FRINGED POLYPORE - *Polyporus arcularius*

**FRUITING BODY:** A small, tough woody polypore with a circular cap, 2-3 cm (¾-1 in); upper surface tan to brown, scaly with delicate white fringe around the periphery; lower surface tan, consisting of many small decurrent pores. **STALK:** Central, tan, 2-4 cm (¾-1½ in) X 2-5 mm (¼ in). **HABITAT:** It is observed singly on dead wood. **EDIBILITY:** Too tough to consider. **COMMENTS:** This is a small, delicate and handsome polypore which is common, occurs in small numbers and can be collected throughout the growing season.

## DYE POLYPORE - *Phaeolus schweinitzii*

**FRUITING BODY:** firm, rosette-shaped, thick, often shelving, 12-20 cm (4½-8 in); upper surface rich brown with yellow margin when young, all becoming dark brown at maturity; lower surface shades of greenish yellow, consisting of many tiny pores.
**STALK:** Not evident. **SPORES:** White to yellow. **HABITAT:** Collected in damp coniferous woods, emerging from the soil at the base of conifers, and firmly attached to buried roots.
**EDIBILITY:** Too tough and possibly poisonous. **COMMENTS:** This fungus causes a disease called red-brown butt rot and can be collected in groups.

GROUP 3

# BLACK-LEG - *Polyporus badius*

**FRUITING BODY:** A tough and woody short-stalked, flat or fan-shaped polypore; 6-15 cm (2-6 in); upper surface shades of brown becoming almost black in the center; lower surface white to tan, consisting of many tiny pores. **STALK:** Central to lateral, tan, 1-5 cm ( ½-2 in) X 1-1.5 cm (½-¾ in) with the lower portion distinctly black. **HABITAT:** This common fungus is collected in large, sometimes shelving groups on downed wood. **EDIBILITY:** Too tough to consider. **COMMENTS:** This large, elegant polypore can be found on dead logs throughout the growing season. It causes a white rot of hardwood and conifer logs.

GROUP 3

## Toothed Fungi

### SCALY-TOOTHED FUNGUS - *Sarcodon imbricatum*

**CAP:** Tough, tan to brown, depressed centrally, 6-10 cm (2-4 in); exhibiting distinct darker brown to black overlapping scales on upper surface; lower surface tan to brown exhibiting teeth or spines, 0.5-0.7 cm (~ ¼) long. **STALK:** Tough, white to tan, 5-7 cm (2-2½ in) X 2-3 cm (1 in). **HABITAT:** It is collected

97

infrequently from soil in wet coniferous woods. **EDIBILITY:** Not recommended. **COMMENTS:** The teeth on the lower surface and scales on the upper surface are distinguishing features. This species of *Sarcodon* is listed as mycorrhizal.

## *Jelly Fungi*

**WITCH'S BUTTER** - *Tremella mesenterica*
**FRUITING BODY:** This fungus has no cap, stalk or gills, but consists of a convoluted or folded growth upon which the spores are formed. The fungus is yellow to orange, 4-6 cm (1½-2½ in) in diameter and has the consistency of firm to flexible jelly when hydrated. **HABITAT:** It occurs on dead wood and is especially noticeable after a rain. **EDIBILITY:** The tough consistency is not appealing. **COMMENTS:** When

dry, this fungus becomes hard, small and inconspicuous. Dried jelly fungi can be rehydrated and after several hours will appear as seen in the photograph.

## BROWN SQUIRREL EARS - *Auricularia auricula*

**FRUITING BODY:** Tan to brown, ear-shaped, rubbery and jelly-like when wet, attached laterally to eccentrically, 4-7 cm (2-3 in). **HABITAT:** It grows on deciduous and coniferous wood, and was collected on bur oak in the Hills. **EDIBILITY:** Listed as edible and good, but has a tough consistency. **COMMENTS:** This fungus has been collected only on oak from one site in the Black Hills. It is much more common in the midwest. Spores are produced on the inner surface of the ear.

## Ridged Fungi

### PIG'S EARS - *Gomphus clavatus*

**FRUITING BODY:** Tan to purplish, broadly club-shaped, expanding near apex; 4-8 cm (1 ½-3 in) wide X 8-12 cm (3-4 ½ in) tall; upper surface centrally depressed; lower surface ridged or veined. **HABITAT:** Fruiting bodies are fused to others in a small group, on soil in wet coniferous woods. We collected it only once. **EDIBILITY:** Listed as edible, but be certain of identification. We think it might be tough. **COMMENTS:** Spores are produced on the veined lower surface. The very edible and delicious chantherelles have a similar structure, but are yellow, gold or orange and are unlikely to be confused with Pig's Ears.

## Resupinate Fungi

**WARTS ON WOOD** - *Peniophora rufa*
**FRUITING BODY:** Pink to pinkish orange, tough, leathery, wrinkled warts, ½ cm (¼ in) in diameter. **HABITAT:** It can be found growing in large groups on dead aspen. **EDIBILITY:** Not worth it. Not recommended. **COMMENTS:** This fungus is observed exclusively on downed aspen in the hills. It is obvious when hydrated so is usually collected after a rain. It causes a white rot of dead aspen.

## Coral and Club Shaped Fungi

### COMMON CORAL - *Ramaria apiculata*

**FRUITING BODY:** Highly compact, rising from a poorly developed stalk, buffy tan, 7-9 cm (3-3½ in) wide X 8-10 cm (3-4 in) tall; mostly parallel branches tapering at tips, sometimes with green coloration. **SPORES:** Yellow, smooth to rough, 7-10 X 3.5-5.5 µm. **HABITAT:** It is found primarily on dead, decomposed conifer wood. **EDIBILITY:** Not recommended. **COMMENTS:** This is one of the most common coral fungi collected mid to late summer. Another very similar species is *R. stricta*, which is more common on deciduous wood. The two species can best be distinguished using microscopic characters. Some species of *Ramaria* may be mycorrhizal.

## TRUNCATE CLUB CORAL - *Clavariadelphus truncatus*

**FRUITING BODY:** Firm, yellow to orange, smooth, club-shaped, 1-3 cm (½-1 in) wide X 8-12 cm (3-4½ in) tall. **SPORES:** Yellow, smooth, 9-13 X 5-8 μm. **HABITAT:** It is collected singly on soil and in groups in coniferous forests in the Hills. **EDIBILITY:** Not considered poisonous, but not recommended because of the texture. However, some report a pleasant taste. **COMMENTS:** This club fungus is quite common during the latter part of the summer. Yellow spores are produced on the surface and sometimes give it a powdery appearance. Another less common club fungus is *C. ligula*, which is slimmer, smaller and not flattened on the top. It usually has a blunt point on top and is cream to buff.

# GROUP 4 - Key to Puffballs, Earth Stars, Stinkhorns and Bird's Nest Fungi in Field Guide

**1A.** Fruiting body is partially (approximately one-half) buried in the soil ............................................................. 2
**1B.** Fruiting body is not buried ................................................. 3
**2A.** Fruiting body is hoof-shaped, outer layer soft and breaking away, sterile base deeply rooted ..................
................Horse Hoof Fungus *(Pisolithus tinctorius)* p. 105
**2B.** Fruiting body rounder, outer layer thick and hard, usually with a small sterile base not deeply rooted ....
................Hard Puffball *(Scleroderma citrinum)* p. 106
**3A.** Fruiting body with a distinct stalk .........................................
................Common Stinkhorn *(Phallus impudicus)* p. 107
**3B.** Fruiting body with no stalk ................................................ 4
**4A.** Fruiting body greater than 1 in. in diameter .................. 5
**4B.** Fruiting body less than 1 in. in diameter ....................... 9
**5A.** Fruiting body round, sculptured, very large, typically 5-12 in., no ostiole ............................................................
................Giant Western Puffball *(Calvatia booniana)* p. 108
**5B.** Fruiting body typically less than 5 in. in diameter ......... 6
**6A.** Outer layer of fruiting body splitting and reflexed ......... 7
**6B.** Outer layer of fruiting body not splitting, but entire ..... 8
**7A.** Outer layer always reflexed ..................................................
................Stalked Earth Star *(Geastrum quadrifidum)* p. 109
**7B.** Outer layer reflexing only when wet, curving inward to enclose the inner spore case when dry .........................
................Water Measurer *(Astraeus hygrometricus)* p. 110
**8A.** Fruiting body white-tan, globose, surface rough with spines, usually on soil .....................................................
................Common Spiny Puffball *(Lycoperdon perlatum)* p. 111
................Peeling Puffball *(Lycoperdon marginatum)* p. 112
**8B.** Fruiting body white-tan, pear-shaped, smooth surface, usually on wood ............................................................
................Pear-Shaped Puffball *(Lycoperdon pyriforme)* p. 113

**9A.** Fruiting body urn-shaped, inner surface of the "cup" striated, open at the top exposing gray "eggs"
......... Striated Bird's Nest Fungus *(Cyathus striatus)* p. 114
**9B.** Fruiting body like a nest and not striated, open at top exposing tan "eggs"
......... Common Bird's Nest Fungus *(Crucibulum laeve)* p. 115

## GROUP 4 - *Puffballs, Earth Stars, Stinkhorns and Bird's Nest Fungi*

### HORSE HOOF FUNGUS - *Pisolithus tinctorius*

**FRUITING BODY:** Dark brown single or double lobed, partially buried in the soil, 5-14 cm (2-6 in) wide by 7-9 cm (2 ¾-3 ½ in) tall. Immature specimens may have some metallic color.
**SPORES:** When young, if the fruiting body is cut longitudinally small seed-like structures (peridioles) are observed and enclose the spores. As the fungus matures the peridioles disintegrate so the entire fruiting body is filled with dark brown, round, spiny spores (7-12 μm). **HABITAT:** Reported in dry, sandy

habitats. Collections in this study are from coal mining spoils. **EDIBILITY:** Definitely disgusting. **COMMENTS:** This ugly fungus is sometimes reminiscent of horse manure. Although its appearance is less than appealing, it is a very important mycorrhizal fungus and has a relationship with pine. In habitats such as coal mining spoils, nutrients for plant growth are sparse and this fungus permits reestablishment of pines. It has the specific epithet *tinctorius* because young parts of the fruiting body have metallic colors that can be used as a dye.

## HARD PUFFBALL - *Scleroderma citrinum*

**FRUITING BODY:** Round or broader than tall, cream to tan, 7-9 cm (3-3 ½ in), wall thick, hard and brittle, may be partially buried; with aging the surface may appear cracked or scaly. **SPORES:** Dark purple-brown at maturity, round, spiny and reticulate, 7-12 µm. **HABITAT:** It is usually collected in dry sandy soil in coniferous/deciduous woods. **EDIBILITY:** Most species of *Scleroderma* are poisonous so no species should be eaten. **COMMENTS:** This species is very similar to another common species, *S. cepa* and can be separated from this species only by the pattern of spore walls. Some species of *Scleroderma* can grow in a mycorrhizal relationship.

## COMMON STINKHORN - *Phallus impudicus*

**FRUITING BODY:** Upper portion 3-4 cm (1-1 ½ in) wide X 4-5 cm (1 ½-2 in) tall, with longitudinal ridges, the areas between ridges black and slimy containing the spores. **STALK:** White and spongy, emerging from an "egg" with remnants remaining at base, 8-10 cm (3-4 in) X 3-4 cm (1-1 ½ in). **SPORES:** Dark brown, smooth, 3-5 X 1.5-2.5 µm. **HABITAT:** Stinkhorns are collected on soil in a variety of habitats, along roads, in grass, and margins of woods. **EDIBILITY:** Can't imagine! **COMMENTS:** The common name of this fungus reflects the strong disagreeable odor described as resembling rotting flesh. This strong smell attracts flies to the black, slimy portion containing spores and flies disperse the spores. It is common some years, especially in the fall. The "eggs", which are near the soil surface break open as the fruiting body emerges.

## GIANT WESTERN PUFFBALL - *Calvatia booniana*

**FRUITING BODY:** White to light tan at maturity, round, soft, and very large, 10-40 cm (4-16 in), attached directly to the soil by a central base; interior white and soft when immature changing to olive brown at maturity and completely filling with dry, powdery masses of spores; outer wall breaking away to expose and release spores. **SPORES:** Olive brown, smooth to spiny, almost round, 3.5-6 μm. **HABITAT:** It is collected in grassy areas in and at edges of woods or on open rangeland. **EDIBILITY:** Puffballs are listed as edible and good, but only when young and white throughout. However, the texture is soft. **COMMENTS:** The large size of this species is the best character to use for identification. It is very important when identifying a puffball to cut it lengthwise to expose the interior, which should be entirely white with no spore formation and which should NOT contain a small immature mushroom. Check for any evidence of a stalk, which would indicate an immature mushroom and probably a poisonous one. There are two other smaller species of *Calvatia* that are commonly collected. *C. cyathiformis* has a purplish-brown spore mass when mature and the spore mass of *C. craniiformis* is yellow-brown. Both of these species have a lower portion that is firm and that does not contain spores.

## STALKED EARTH STAR - *Geastrum quadrifidum*

**FRUITING BODY:** Inner spore-containing structure is round, soft and brown, 2-3 cm (¾-1 in), with an ostiole (central opening), located on a pedicel (little stalk) and located inside outer, firmer walls, which are split into several reflexing rays; the entire fruiting body measuring 5-7 cm (2-3 in). **SPORES:** Brown, round, rough, 3.5-5 µm. **HABITAT:** It is found on soil or debris in mixed coniferous/deciduous woods. **EDIBILITY:** Inedible. **COMMENTS:** The inner soft spore case contains many dry spores, which exit through the opening when the case is touched. The split reflexed outer walls gives this fungus its common name. The specimen in the photo is weathered. The spore case is bleached a lighter color and the distinct lip around the ostiole eroded. Another species of earth star collected in the Black Hills is *G. saccatum*, which is distinguished from the above by a spore case which is not on a pedicel. *G. triplex*, another common species of earth star displays a spore case sitting in a saucer-like structure.

*Geastrum triplex*

## WATER MEASURER - *Astraeus hygrometricus*

GROUP 4

**FRUITING BODY:** Inner spore-containing structure is soft, tan to gray, 1.5 -2.5 cm (½-1 in) in diameter. Darker brown, outer firmer walls are split and reflexed outwardly when wet forming rays, but enclose the inner spore-containing structure when dry. When extended the entire fruiting body is 5-6 cm (2-2 ½ in) in diameter. **SPORES:** Brown, round and rough, 7-11 μm. **HABITAT:** This fungus is collected primarily in dry, sandy habitats with conifers. **EDIBILITY:** Not recommended. **COMMENTS:** This fungus can be easily confused with the true earth stars. The significant difference is that the rays are hygroscopic, extending outward and flattening when wet and curling up around the inner spore case when dry. It is easy to separate this fungus from an Earth Star by placing the Water Measurer in water and watching the rays extend.


**COMMON SPINY PUFFBALL -** *Lycoperdon perlatum*
**FRUITING BODY:** Soft, white to shades of brown, covered with conical-shaped spines, 4-6 cm (1 ½-2 in) wide X 5-8 cm (2-3 in) tall, opening by an ostiole. **SPORES:** Brown, round and minutely spiny, 3.5-4.5 μm. **HABITAT:** It is commonly collected in groups on soil. **EDIBILITY:** It is listed as edible

111

GROUP 4

when young, firm and the contents are white. Always cut it longitudinally to examine contents before eating. Exercise the same caution as with the Giant Western Puffball because it can be confused with immature poisonous species of *Amanita*.
**COMMENTS:** This puffball can also be pear-shaped, but the spines and soil habitat are useful characters for identification.

**PEELING PUFFBALL** - *Lycoperdon marginatum*
**FRUITING BODY:** Soft, white to shades of tan and brown, covered with spines that converge at the tips forming pyramid-shaped structures when mature, outer layer with spines peeling off in sheets exposing a softer inner layer, 3-5 cm (1-2 in).
**SPORES:** Olive/brown, round and spiny, 3-5 µm. **HABITAT:** These puffballs are collected primarily in dry sandy soils.
**EDIBILITY:** It is listed as edible when young, firm and the

GROUP 4

contents are white. Use the same caution as described for the Common Spiny Puffball. **COMMENTS:** The spines converging at the tips and the manner in which the outer layer peels away are good characters to use for identification. You may need a hand lens to observe the spines. The specimen shown is immature.

**PEAR-SHAPED PUFFBALL** - *Lycoperdon pyriforme*
**FRUITING BODY:** Smooth, soft, white to shades of brown, pear-shaped, 2-3 cm (¾-1 in) wide X 2-3 cm (¾-1 in) tall, opening by an ostiole; slightly smaller than the Common Puffball. When immature the inside of the rounded upper portion is white with a marshmallow texture becoming olive-brown with spores at maturity. Below the upper portion is the sterile base, which does not contain spores. **SPORES:** Olive-brown, round and smooth, 3-4.5 μm. **HABITAT:** A common fungus growing

GROUP 4

in groups on dead wood. **EDIBILITY:** It is listed as edible, but caution must be used because it can be confused with poisonous immature buttons of *Amanita*. Always cut the fruiting body longitudinally and examine it for structures that resemble a young mushroom. If these are present **DO NOT EAT**. It is important to choose only immature, young puffballs with no spore development for eating. **COMMENTS:** This puffball usually occurs on wood, which is a useful character to separate it from the Common Spiny Puffball.

**STRIATED BIRD'S NEST FUNGUS** - *Cyathus striatus*
**FRUITING BODY:** Resembles a small bird's nest at maturity, 1-1.5 cm (½-¾ in). Outside layer is gray and hairy with several gray, shiny "eggs" inside that contain the spores. The inside of the outer layer is striated. The "nest" is covered with a white structure in immature specimens. **SPORES:** colorless, smooth, 22-40 X 18-30 µm. **HABITAT:** It is collected in groups on dead

**GROUP 4**

wood and other vegetation. **EDIBILITY:** Too small to consider.
**COMMENTS:** Dispersal is facilitated by raindrops which causes the "eggs" containing the spores to shoot outside the cup.

**COMMON BIRD'S NEST FUNGUS** - *Crucibulum laeve*
**FRUITING BODY:** Outer surface tan to brown, at maturity resembling a tiny bird's nest, 0.5-1.0 cm (¼ to ½ in). The inside contains several tan "eggs", which contain the spores. "Eggs" are attached by a cord to the outer layer. The "nest" will be covered with a white membrane on young specimens.
**SPORES:** Smooth, 7-10 X 3.6 µm. **HABITAT:** Usually collected in groups on dead wood and other vegetation. **EDIBILITY:**

GROUP 4

Too small to merit eating. **COMMENTS:** The small size makes this unique fungus easy to miss. Dispersal of "eggs" is initiated by raindrops forcibly shooting them to adjacent vegetation where they attach by a sticky substance.

# GROUP 5 - *Key to Cup Fungi in Field Guide*

**1A.** Fruiting body cup-shaped ............................................................ 2
**1B.** Fruiting body not a simple cup-shape, but has ridges
and pits or is convoluted or saddle-shaped ............................... 5
**2A.** Fruiting body a shallow cup, tan-brown, rough and
wrinkled on inside and outside surfaces, attached
centrally at the bottom ...................................................................
................................ Convoluted Cup Fungus *(Discina perlata)* p. 118
**2B.** Fruiting body a shallow cup, various colors and not rough
and wrinkled inside .................................................................... 3
**3A.** Fruiting body tan-brown, inside and outside surfaces
smooth, attached centrally at the bottom .................................
.................................... Smooth Cup Fungus *(Peziza repanda)* p. 119
**3B.** Fruiting body colors other than tan-brown ......................... 4
**4A.** Small orange cups, white and floccose on outer surface
.................... False Orange Peel Cup *(Sowerbyella rhenana)* p. 120
**4B.** Small orange-red cups, brown bristles around periphery
........................................ Eyelash Cup *(Scutellinia scutellata)* p. 121
**5A.** Upper portion of fruiting body consisting of many pits,
surrounded by ridges; stalk present ................................... 6
**5B.** Upper portion of fruiting body convoluted or
saddle-shaped, without pits and ridges; stalk present ....... 7
**6A.** Ridges around pits black .........................................................
.................................... Black Morel *(Morchella angusticeps)* p. 122
**6B.** Ridges same color as pits .........................................................
................................................ Tan Morel *(Morchella esculenta)* p. 123
**7A.** Fruiting body tan to russet brown, almost as wide as tall,
upper portion convoluted and deeply folded; stalk
wrinkled ....... Snowbank False Morel *(Gyromitra gigas)* p. 124
**7B.** Fruiting body taller than wide, upper portion almost
saddled-shaped ..................................................................... 8
**8A.** Upper portion of fruiting body orange-brown, incurved
................ Orange-Saddled False Morel *(Gyromitra infula)* p. 125
**8B.** Upper portion of fruiting body shades of gray/tan ..........
.................... Gray-Saddled False Morel *(Helvella ephippium)* p. 126

# GROUP 5 - *Cup Fungi*

## CONVOLUTED CUP FUNGUS - *Discina perlata*

**FRUITING BODY:** A shallow cup, both upper and lower surfaces tan to brown and both surfaces folded or wrinkled, lower surface lighter colored and attached to a very short stalk or no stalk at all, up to 12 cm (5 in) in diameter with incurved margin, which may divide into lobes. Texture is brittle and fruiting bodies break easily. Spores form on the inner surface of the cup. **SPORES:** Colorless, smooth, spindle-shaped, 25-35 X 11-16 µm, with one large central oil drop and a smaller oil drop at each end, 8 spores/ascus. **HABITAT:** It occurs on the ground or on moss in coniferous woods and is associated with pine. **EDIBILITY:** It is not recommended because it can be confused with other brown cup-shaped fungi. **COMMENTS:** Fruiting bodies usually occur in the spring when moisture is abundant. As the cup dries, the spores are shot into the air and appear as smoke trails rising from the cup. This fungus can be confused with *Peziza repanda*, described below, which is less wrinkled. Microscopic examination of the spores is the best method to distinguish the two fungi.

## SMOOTH CUP FUNGUS - *Peziza repanda*

**FRUITING BODY:** Size, texture and color of the cup are similar to the Convoluted Cup described on the previous page, but upper and lower surfaces not heavily wrinkled, 4-12 cm (1 ½-5 in). **SPORES:** Formed on inner surface of cup, colorless, smooth, 14-18 X 8-10 µm, 8 spores/ascus. **HABITAT:** This cup

*Peziza badia*

fungus is found on soil, moss or wood in coniferous and deciduous woods. **EDIBILITY:** It is not edible. **COMMENTS:** The fungus appears in the spring and commonly throughout the summer when moisture is available. Another less common cup fungus is *Peziza badia*, which is smaller, less spreading and has a darker reddish-brown lower surface.

## FALSE ORANGE PEEL CUP - *Sowerbyella rhenana* (=*Aleuria rhenana*)

GROUP 5

**FRUITING BODY:** Smooth, gold to orange cups, inner surface brighter orange than outer surface, which has a whitish, mealy appearance, 1.5-2 cm (½-¾ in); cups on short stalks. **SPORES:** Formed on inner surface of cup, colorless, reticulate, 20-23 X 11-13 µm, 8 spores/ascus. **HABITAT:** It is collected singly growing on moss in wet bog birch/spruce/moss habitats. **EDIBILITY:** Too small to be considered. **COMMENTS:** This small orange cup is rarely collected and apparently uncommon.

## EYELASH CUP - *Scutellinia scutellata*

**FRUITING BODY:** A tiny shallow cup, interior orange, pink to red, exhibiting distinct brown bristles around the periphery, 0.5-1.0 cm (¼-½ in). **SPORES:** Formed on inner surface of cup, colorless, 17-19 X 11-14 µm, 1- several oil drops. **HABITAT:** It occurs in groups growing on very wet, well-decomposed dead wood. **EDIBILITY:** Too small to consider. **COMMENTS:** A very elegant, tiny cup fungus and relatively common in wet woods.

## BLACK MOREL - *Morchella angusticeps*

**FRUITING BODY:** Conical to cylindrical upper portion, 6-8 cm (2 ½-3 in) tall X  3-4 cm (1 X 1 ½ in) wide; consists of a series of black ridges separating numerous tan, elongated pits. **STALK:** Top portion is attached to a tan, hollow stalk, 3-5 cm (1-2 in) tall X 1-2 cm (½-¾ in) wide. **SPORES:**  Formed at bottom of the pits, cream, smooth, 21-24 X 12-14 µm, 8 spores/ascus. **HABITAT:** It is found on soil in mixed  coniferous/deciduous woods during the early spring. **EDIBILITY:** The Black Morel is listed as edible and frequently eaten, but some individuals experience some mild poisoning after eating. **COMMENTS:** This is the most common morel in the Black Hills and is collected frequently in the northern Hills. Mycophagists intensively hunt for it during the spring. It can be confused with False Morels, which also appear during the early spring and which are poisonous. The Black Morel will have an upper portion that is divided into definite and separate pits, compared to some false morels which are extensively folded (See description for the *Gyromitra* species on pages 124-125). The bottom of the cap is attached to the hollow stalk. Cut lengthwise to determine this character.

### TAN MOREL - *Morchella esculenta*

**FRUITING BODY:** Very similar to the black morel, except ridges are tan and not black and it is larger. **STALK:** Similar to the stalk of the black morel. **SPORES:** Formed at bottom of pits, 20-25 X 12-16 µm, 8 spores /ascus. **HABITAT:** It is collected on soil in mixed coniferous/deciduous woods and prairies during early spring. **EDIBILITY:** Edible and delicious. **COMMENTS:** It has the same ridge/pit and stalk characters as the Black

Morel. It is less common than the Black Morel and observed more in the southern rather than northern Hills and at lower elevations.

## SNOWBANK FALSE MOREL - *Gyromitra gigas*

**FRUITING BODY:** Tan-russet brown upper portion is deeply wrinkled and convoluted, but showing no separate pits; cap 7-10 cm (2¾-4 in) tall and wide. **STALK:** White to tan, 5-7cm

(2-3 in) tall and wide. **SPORES:** Formed on outside of convoluted cap, 24-36 X 10-15 µm, one central oil drop and smaller ones at each end, 8 spores/ascus. **HABITAT:** Occurs on soil in mixed coniferous/deciduous woods of pine, spruce and aspen during early spring. **EDIBILITY:** Not recommended. Since many False Morels are **deadly poisonous**, none should be eaten. **COMMENTS:** This heavy, robust fungus is called the Snowbank False Morel because the fruiting body emerges very early in the spring, often just at the edge of a melting snowbank and at about the same time as true morels. Some collectors mistake it for the edible true morels.

## ORANGE-SADDLED FALSE MOREL - *Gyromitra infula*

GROUP 5

**FRUITING BODY:** Orange upper portion saddle-shaped to lobed, margin usually incurved, 5-7 cm (2-3 in) wide X 4-6 cm (1 ½-2 in) tall. **STALK:** White to tan, indistinctly furrowed, 4-6 cm (1 ½-2 in) X 1-2 cm (½-¾ in). **SPORES:** Formed on outer surface of convoluted cap, colorless, 17-23 X 7-10 µm, 2 large oil drops, 8 spores/ascus. **HABITAT:** It is infrequently found growing solitary on dead wood or soil in damp woods. **EDIBILITY: Poisonous and can cause cancer. COMMENTS:** This fungus is called a false morel because some collectors confuse it with the edible true morels. Distinguishing features of *G. infula* are the orange saddle-shaped to lobed upper portion. It does not have the ridge/pit system of a true morel.

## GREY-SADDLED FALSE MOREL - *Helvella ephippium*

**FRUITING BODY:** Upper portion grayish/tan, smooth and saddled-shaped, downy beneath, 3-4 cm (1-1½ in). **STALK:** Cream to tan/gray, downy, slightly enlarged at base,

4-5 cm (1½-2 in) X 0.5 cm (¼ in). **SPORES:** Formed on outside of saddle-shaped cap, smooth with large central oil drop, 17-18 X 11.0-12.5 µm, 8 spores/ascus. **HABITAT:** This small fungus is collected in small numbers in moss in wet mixed coniferous/deciduous woods. **EDIBILITY:** Unknown and too small and fragile to consider. **COMMENTS:** It is an unusual fungus, observed only twice during the study. We have identified our collections as *H. ephippium*, but the taxonomy is problematic.

## GROUP 6 - *Parasitic Fungi*

The field guide includes a few parasitic fungi because they are encountered in the Black Hills. The first three parasitic fungi shown here are related to the cup fungi, discussed previously, but instead of having visible fruiting bodies, their fruiting bodies are microscopic and observed only by what appears as a small dot. Each dot represents a tiny hole or ostiole into a flask-shaped structure or urn (perithecia) in which spores are formed. The examples included here form their perithecia in a mass of firm mycelium called a stroma.

### BLACK KNOT - *Dibotryon morbosum*

### GROUP 6

**FRUITING BODY:** It appears as a hard, black tumor-like growth on chokecherry branches that typically reaches 15 cm (6 in) long. This black tumor-like growth is called a stroma and embedded in this stroma are myriads of tiny, microscopic flask-shaped fruiting bodies called perithecia, in which the spores are formed. If you look carefully you may see myriads of tiny black dots, which are the openings (ostioles) through which the spores disperse. **HABITAT:** It is found growing primarily on chokecherry branches. **EDIBILITY:** Certainly not. **COMMENTS:** This fungus is a parasite on chokecherry and areas of branches distal to the black tumor-like growths will eventually die. Heavily infected shrubs will slowly die. Black Knot is so common in the Hills and so specific for chokecherry that you can identify chokecherry by the Black Knot.

## LOBSTER MUSHROOM - *Hypomyces lactifluorum*

**CAP:** Of host mushroom is a bright orange, centrally-depressed surface, 7-9 cm (2¾-3 ½ in) wide. **GILLS:** Appear not as true gills, but as shallow ridges. Cap and gills are covered by an orange stroma produced by the parasite that contains numerous tiny dark orange dots. **STALK:** Of host mushroom stout, 3-4 cm (1-1 ½ in) wide and tall. **SPORES:** White. **HABITAT:** Occurs in mixed pine and aspen habitats where the host mushroom would be located. **EDIBILITY:** A hard fungus and not recommended. **COMMENTS:** This unusual fungus is really covering a parasitized and deformed mushroom (probably *Lactarius* or *Russula* species). The parasitic relationship causes the mushroom to deform and gives it a different appearance. *H. lactifluorum* is the parasitic fungus, which forms tiny flask-shaped fruiting bodies all over the mushroom, but most noticeably on what would be the gill area. Spores are produced in these tiny flask-like structures and the dark orange dots are the openings by which the parasite's spores are released. The parasitized mushrooms are a favorite food source for squirrels. It is possible to confuse this fungus with the delicious and edible chantarelle.

## GREEN LOBSTER - *Hypomyces luteovirens*

GROUP 6

**CAP:** Of host mushroom tan to yellow to brown, flat to depressed centrally, 4-6 cm (1 ½-2 ½ in). **GILLS:** Not observed, but replaced by the growth of the parasite, which is a continuous dimpled bright green stroma extending downwards to colonize part of the stalk. **STALK:** Of host mushroom may be bright green, colonized by the parasite or be the color of the cap, 5-6 cm (2-2 ½ in) X 2-3 cm (¾-1 in). **SPORES:** White. **HABITAT:** The parasite occurs in mixed pine and aspen habitats where *Russula* or *Lactarius* species would be growing. **EDIBILITY:** Not known. **COMMENTS:** This unusual fungus is really a parasite that has deformed the mushroom (probably *Lactarius* or *Russula* species). *Hypomyces luteovirens* is the name of the parasitic fungus, which forms very tiny flask-shaped fruiting bodies all over the fungus, but most noticeably on what would be the gill area, destroying the gills. Spores are produced in these tiny flask-like structures and the darker dots are the openings by which the parasite's spores are released. It is easy to miss this one and think you are seeing just another *Russula* until you observe the gill area.


### ERGOT - *Claviceps purpurea*

Ergot is the name of a disease on grasses that replaces the grain with hard, dark sclerotia, which are 2-3 cm (¾-1 in) long. It is caused by a fungus, called *Claviceps purpurea*, which is also related to cup fungi. It usually occurs on grasses that cross pollinate such as rye as seen in the photograph. In our area, it is observed more frequently on smooth brome grass. The sclerotia contain many **toxic** alkaloids which can cause constriction of the blood vessels in the extremities of cattle and sheep that eat the diseased grass. This malady called ergotism is not only painful but can result in loss of ears, tails and hooves. Erogotism can also occur in humans if they eat rye bread contaminated with the sclerotia. In the Middle Ages the disease was called St. Anthony's Fire due to the religious order that cared for suffering patients. The sclerotia

also contain alkaloids that are psychoactive and there is some evidence that ergotism was involved in the infamous witch trials in Salem, Massachusetts in the 1600s.

**PINE GALL RUST** - *Endocronartium harknessii*
The parasite forms woody, oblong to pear-shaped galls or thickened areas on branches of ponderosa pine. It occurs throughout the Black Hills on pine. The rust produces orange spores on the outside of the gall each spring which infect

other pine. Galls grow larger each year until they have girdled and killed the branch. Quality of timber is reduced and heavy infections can slowly kill the tree. Young and older galls are shown. The older gall shown on the right is 6 cm (2 ½") X 4 cm (1 ½") but they can be larger.

## GROUP 7 - *Lichens*

Although lichens are only partially comprised of fungi, a few of the most common ones encountered in the Black Hills have been included in the field guide. Lichens are a mutually beneficial relationship between an alga and a fungus. Although they are often growing on trees, they are not parasites.

## GREEN FAIRY CUPS - *Cladonia* species

**THALLUS:** Small, green, flat and leaf-like, 0.5-1.0 cm (< ½ in).
**FRUITING BODY:** Most of these lichens produce upright, green cup-shaped fruiting bodies, some red or brown at the tips, that measure 1.0-1.5 cm (½ - ¾ in). Some species have fruiting bodies more club-shaped than cup-shaped. **HABITAT:** They are encountered growing on soil, wood, or moss in forested areas. **EDIBILITY:** Reported as a good food source for animals, but not recommended for human consumption. **COMMENTS:** This is a very common lichen in the Hills. The upright character of fruiting bodies places it in the fruticose group of lichens. Species of *Cladonia* and other genera of lichens are a major component of the biota in the arctic and are a major food source for caribou.

## OLD MAN'S BEARD - *Usnea cavernosa*

**THALLUS:** Fruticose, gray-green pendulous or stringy growths draped on branches of conifers, 20-40 cm (8-16 in). **FRUITING BODY:** Not observed. **HABITAT:** It is found growing primarily on conifers in coniferous forests. **EDIBILITY:** It is reported as a food source for animals, but not recommended for human consumption. **COMMENTS:** It is very common, primarily growing upon pines and spruce. Although it attaches to branches, it is not parasitic because the algal component is photosynthetic, providing the food source. Another species of *Usnea* also very common in the Hills on branches of conifers is *U. hirta*. The thallus of *U. hirta* is more condensed and less pendulous with many short branches. Apparently deer eat these lichens and during the winter this lichen can be an important part of their diet.

## SUNBURST ORANGE LICHEN - *Xanthoria montana*

**THALLUS:** Foliose, yellow to orange, small, flat and leaf-like with rhizines, 3-5 mm (<1/4 in). **FRUITING BODY:** Darker orange, raised, shallow cups, 5 mm (<1/4 in). Ascospores average 13.3 X 5.3 µm. **HABITAT:** Found growing on primarily deciduous wood. **EDIBILITY:** Not recommended. **COMMENTS:** This is a very common orange lichen on wood. *Xanthoria polycarpa* and *X. hasseana* are similar, but *X. polycarpa* does not have rhizines and ascospores of *X. hasseana* are larger. Another species of *Xanthoria*, also orange and common on wood is *X. fallax*, which produces occasional cup-shaped fruiting bodies.

## VEINED DOG LICHEN - *Peltigera canina*

**THALLUS:** Foliose, gray to brownish green, flat and leaf-like, 1.5-2.5 cm (½-1 in), white to tan lower surface with shallow veins and root-like structures, called rhizines. **FRUITING BODY:** Incomplete, raised, brown cups may or may not be present. **HABITAT:** Growing on soil, moss, wood or rocks in forested habitats. **EDIBILITY:** Not recommended. **COMMENTS:** This is one of the most common lichens in the hills. It can be confused with liverworts, which are non-vascular plants. Another less common species of *Peltigera* is *P. aphthosa*, which is greener and has dark green to black warts on the upper surface of thallus.

# GROUP 8 - *Slime Molds*

## WHITE CORAL SLIME - *Ceratiomyxa fruticulosa*

**FRUITING BODY:** White, glistening, very soft and fragile, consisting of numerous small and fine stalks, producing a mass that appears like coral, 1 cm (½ in) X 1 cm (½ in). **STALK:** None, attached directly on the wood. **SPORES:** Colorless, smooth, 10-13 X 6-7 µm. **HABITAT:** It grows in groups on very wet, well-decomposed wood. **EDIBILITY:** Too small and ephemeral to be of value. **COMMENTS:** This is a rather uncommon slime mold occurring in damp, shaded woods. It is difficult to collect because it collapses when touched.

## HERDS OF WHITE CRUST - *Physarum diderma*

**FRUITING BODY:** Clustered, spherical to globose, outer wall white and calcareous or chalky, approximately 1.0 mm (< 1/8 in) in diameter. Outer wall breaks easily and reveals a yellowish network (capillitia) and when filled with spores the inside will appear black. **STALK:** Little or no stalk present. **SPORES:** Black and powdery, rough, almost round, 10-12 μm. **HABITAT:** This tiny slime mold is found in masses early in the spring on dead vegetation. **EDIBILITY:** Definitely too small and ephemeral to be of value. **COMMENTS:** Early in the spring shortly after snowmelt, it is possible to find large groups of this tiny slime mold. The fruiting bodies are close together and sometimes appear fused.

## CHOCOLATE TUBES - *Stemonitis fusca*

**FRUITING BODY:** Chocolate brown, cylindrical, spongy-appearing, 5-10 mm (1/4 in) tall X 1 mm (< 1/8 in) wide. **STALK:** Very thin, shiny, dark brown to black, appearing as a wire running vertically through the fruiting body. **SPORES:** Brown, minutely spiny, 7.5-9 μm. **HABITAT:** Located in groups on wet, decomposed logs. **EDIBILITY:** Too small and fragile to be of value. **COMMENTS:** Chocolate Tubes are relatively common

# GROUP 8

on wet logs in damp areas. Fruiting bodies are very small and can be easily missed unless they are closely observed. Each fruiting body appears as a very tiny, cylindrical brown tree.

## SCRAMBLED EGG SLIME - *Fuligo septica*
**FRUITING BODY:** Appears as a white (sometimes with shades of yellow) crusty or powdery, cushion-shaped growth, 2-9 cm (1 ½-3 ½ in), enclosing dark brown to black spores. **SPORES:** Brown to black, minutely spiny, 6-9 µm. **HABITAT:** It grows upon a variety of dead vegetation. Common and collected frequently under Rocky Mountain junipers in the southern hills as well as other locations in the area. **EDIBILITY:** Not appealing for consumption. **COMMENTS:** This is the most common slime

# GROUP 8

mold seen in the Hills, probably because it is one of the largest in size. It has been described as a "white blob" and is the only species that is not aesthetically appealing. The white covering is composed of granular lime. A very similar species and collected only from the southern hills is *Fuligo megaspora,* which has dark, very rough spores, 15-20 μm in diameter. It can be distinguished from *F. septica* only by spore size and shape.

**PINK AND TAN BUBBLES** - *Lycogala epidendrum*
**FRUITING BODY:** Smooth, globose, 4-10 mm (~ ¼ in), wet, soft and pink when immature becoming dry and tan at maturity, containing pinkish/tan dry spores. **SPORES:** Pinkish to tan, reticulate, 6-7.5 μm. **HABITAT:** It is seen on dead, decomposing

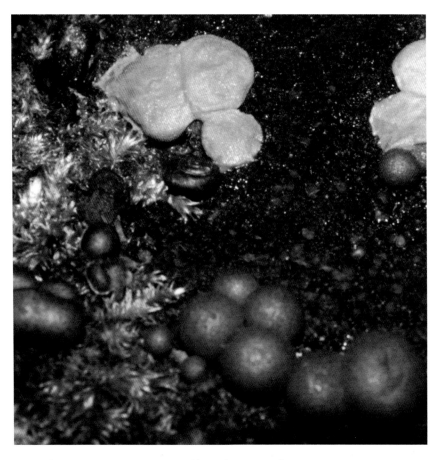

wood. **EDIBILITY:** Too small and insignificant. **COMMENTS:** A very common slime mold. The immature fruiting bodies of this slime mold have been described as resembling pink bubble gum. Mature fruiting bodies are small and can be easily missed because they may be the same color as the wood.

# Mushroom Calendar of Expected Collection Times

Collection times are shown for selected mushrooms and other fungi. Months are approximate because time of fruiting can be extremely variable from year to year. Times shown are based on when these fungi were collected during a five-year period. The woody brackets or shelf fungi (polypores) persist throughout the year and can be collected nearly anytime. Consequently, they have been not included in the calendar.

| Fungus Name | May | June | July | Aug | Sept | Oct |
|---|---|---|---|---|---|---|
| *Agaricus* spp. [1] | ■ | | | ■ | ■ | ■ |
| *Amanita* spp. | | ■ | ■ | ■ | ■ | |
| *Armillaria ostoyae* | | | | | ■ | ■ |
| *Calvatia* spp. | ■ | | | ■ | ■ | |
| *Clitocybe gibba* | | | ■ | | | |
| *Coprinus* spp. | ■ | | | ■ | ■ | ■ |
| *Cortinarius* spp. | | | | ■ | ■ | |
| *Geastrum* spp. | ■ | | | | ■ | |
| *Gymnopilus* spp. | | | | ■ | ■ | |
| *Gymnopus dryophilus* | | ■ | ■ | ■ | | |
| *Hygrophoropsis aurantiaca* | | ■ | | | | ■ |
| *Hygrophorus* spp. | ■ | | | ■ | ■ | |
| *Inocybe fastigiata* | ■ | | ■ | ■ | | |
| *Lactarius* spp. | | | | ■ | ■ | |
| *Leccinum insigne* | | | ■ | ■ | | |
| *Lepiota* spp. | | | | ■ | ■ | ■ |
| *Leucopaxillus albissimus* | ■ | | | ■ | ■ | |
| *Lycoperdon* spp. | ■ | | ■ | ■ | ■ | ■ |
| *Morchella* spp. | ■ | ■ | | | | |
| *Mycena* spp. | ■ | ■ | | ■ | ■ | ■ |
| *Panaeolus* spp. | ■ | ■ | ■ | ■ | ■ | |
| *Peziza repanda* | ■ | ■ | ■ | | ■ | |
| *Pholiota* spp. | ■ | | ■ | ■ | ■ | |
| *Pleurotus populinus* | ■ | | ■ | | | |
| *Pluteus cervinus* | | ■ | ■ | ■ | | ■ |

| Fungus Name | MAY | JUNE | JULY | AUG | SEPT | OCT |
|---|---|---|---|---|---|---|
| *Psathyrella candolleana* | ■ |  | ■ | ■ |  | ■ |
| *Ramaria* spp. |  | ■ | ■ |  |  |  |
| *Rhodotus palmatus* |  | ■ | ■ |  | ■ |  |
| *Russula* spp. |  | ■ | ■ |  | ■ |  |
| *Scutellinia scutellata* |  | ■ | ■ | ■ |  |  |
| *Suillus granulatus* | ■ | ■ |  |  |  |  |
| *Tremella mesenterica* | ■ | ■ | ■ |  | ■ | ■ |
| *Tricholomopsis* spp. |  | ■ | ■ |  |  |  |
| *Xeromphalina campanella* | ■ | ■ | ■ | ■ |  |  |

¹ = spp. indicates more than one species

# GLOSSARY

## A

**ALGAE** — microscopic photosynthetic organisms, some of which form a mutualistic or beneficial relationship with some fungi to form lichens.

**ANNUAL** — grows only one year and then dies.

**ANNULUS** — a ring around the stalk or stipe of some mushrooms; partial remnants of a veil that cover the gills.

**ASCUS** — a sac-like structure or cell found in the cup fungi in which the spores or ascospores are located. Plural = asci.

## B

**BASIDIA** — special cells on which basidiospores are formed.

**BIRD'S NEST FUNGI** — a group of fungi that resemble tiny bird's nests; spores are enclosed in peridioles ("eggs") in the exoperidium ("nest").

**BOLETES** — a group of fungi that have the form and shape of mushrooms, but have pores instead of gills under the cap.

| | |
|---|---|
| BRACKETS | a group of fungi that usually grows on trunks or branches of trees as a shelf or shelves. |
| BRITTLE | refers to gills and stipes that are not soft, but break with a crack. |
| BROWN ROT | a decomposition of wood resulting in brown cubes due to the preferential decay of cellulose over lignin. |
| BULBOUS | describes an expanded portion of the base of a stalk. |

## C

| | |
|---|---|
| CAP | the upper portion of most larger, fleshy fungi; called a pileus when referring to mushrooms. |
| CAPILLITIA | stringy material found in fruiting bodies of slime molds and some puffballs. |
| CLUB FUNGI | a group of fungi that are shaped like a club, or baseball bat. |
| COBWEBBY VEIL | a structure observed on some immature mushrooms that covers the gills and reaches from the edge of the cap to the stalk. |
| CONCENTRIC ZONES | areas on caps of some fungi that are circular or semicircular. |
| CONIFERS | trees that have needles or scales and typically do not drop all of them in the fall; ie. pines, spruce, junipers. |
| CORAL FUNGI | a group of fungi resembling the animals that form the coral reefs in warm shallow seas. |
| CROWDED GILLS | refers to gills close together. See *Leucopaxillus albissimus* on page 46. |
| CRUSTOSE | a growth form of lichens with a flat thallus that adheres very tightly to the substrate so there are no loose edges. |
| CUP | a structure from which some mushrooms emerge; usually below the soil line; the top breaks open to release the emerging mushroom; a volva. |
| CUP FUNGI | a group of fungi which have fruiting bodies that are shaped like cups, or have a ridge/pit system (the pits are shallow cups), saddle-like caps interpreted as cups or convoluted caps, also interpreted as cups. |
| CUTICLE | the outer layer of the pileus or cap of a mushroom. |
| CYSTIDIA | cells larger than spores formed on gills of some mushrooms. |

## D

| | |
|---|---|
| DECIDUOUS | trees that lose their leaves in the autumn. |

| | |
|---|---|
| DECOMPOSER | an organism that can utilize dead organic substrates for growth and returns nitrogen, phosphorus, potassium and other nutrients to the soil; primarily fungi and bacteria. |
| DELIQUESCE | refers to a few fungi that produce enzymes that digest their fruiting body. |
| DELIRIUM | confused and abnormal mental state. |
| DICHOTOMOUS | refers to gills of some mushrooms that split into two gills between the stalk and edge of the cap. |
| DISTANT GILLS | refers to gills that are well-spaced. See *Xeromphalina campanella* on page 37. |

### E

| | |
|---|---|
| EARTH STAR | a puffball with an exoperidium (outer layer) that splits and extends to resemble points of a star and an endoperidium (inner layer) that encloses the spores. |
| ECCENTRIC | not central or symmetric; to the side, lateral. |
| ENDOPERIDIUM | the inner layer of puffballs, earth stars and bird's nest fungi containing the spores. |
| EPHEMERAL | refers to a structure that lasts a short time, ie. some rings on the stalk of mushrooms disappear. |
| EXOPERIDIUM | the outer layer of puffballs, earthstars and bird's nest fungi. |

### F

| | |
|---|---|
| FRUITING BODY | the sexually reproductive portion of a fungal life cycle; the structure that produces spores; typically the visible portion of the life cycle. |
| FRUTICOSE | a growth form of lichens with a thallus that is hanging or growing upright. |
| FOLIOSE | a growth form of lichens with a flat and leaf-like thallus and edges that are free of the substrate. |
| FUNGI | a group of organisms comprised of filaments (mycelia) that feed as saprobes, parasites or form a mutualistic relationship; need moisture but no sunlight to grow; some need light to fruit or complete a life cycle. |

### G

| | |
|---|---|
| GILLS | the plate-like structures on the lower surface of the cap of mushrooms; the location of spore production. |

### H

| | |
|---|---|
| HALLUCINOGEN | a substance that elicits unrealistic perceptions; causes a disorder of the nervous system. |

| | |
|---|---|
| **HARDWOODS** | typically refers to trees other than conifers; those that lose all their leaves in the autumn; deciduous. |
| **HYALINE** | colorless, commonly used to describe spore color. |
| **HYGROPHANOUS** | usually refers to caps of mushrooms which change color as they dry; appear water-soaked when moist. |
| **HYGROSCOPIC** | refers to some fungi with parts that move and change shape in response to moisture; ie. some fungi have rays that open when wet and close when dry. |
| **HYPOGEOUS** | grows underground. |
| **HYPHA (HYPHAE)** | the filamentous portion of a fungus representing the structural entity. |

## J

| | |
|---|---|
| **JELLY FUNGI** | fungi with a fruiting body that is gelatinous when moist and hard when dry; can be rehydrated to become gelatinous again. |

## K

| | |
|---|---|
| **KEYS** | sets of statements usually in couplets that utilize characteristics of organisms to assist in identifying them. |

## L

| | |
|---|---|
| **LAMELLAE** | the plate-like structures on the lower surface of the cap of mushrooms; the location of spore production. |
| **LATEX** | a liquid substance produced by some mushrooms; ie. species in the genus *Lactarius*. |
| **LICHEN** | a thallus resulting from a mutualistic beneficial relationship between an alga and a fungus. |

## M

| | |
|---|---|
| **MORPHOLOGY** | the structure or appearance that includes shape, size and color of a fungus. |
| **MUSHROOMS** | fleshy fungi that have a fruiting body with a cap (pileus), gills (lamellae) and stalk (stipe). |
| **MUTUALISTIC** | an intimate relationship between two different organisms in which both benefit; ie. lichens (algae and fungi) and mycorrhizae (plants and fungi). |
| **MYCELIUM** | a mass of hyphae. |
| **MYCOLOGY** | the study of fungi. |
| **MYCOLOGIST** | an individual who studies fungi. |
| **MYCOPHAGISTS** | individuals that eat mushrooms. |
| **MYCORRHIZAL** | a beneficial relationship between many fungi and roots of plants in which the plant provides the carbon |

source (food) for the fungus and the fungus absorbs the needed nutrients (nitrogen, potassium, phosphorus) for the plants.

## O

| | |
|---|---|
| OSTIOLE | an opening in some fruiting bodies for spore release; also an opening to a flask-shaped fruiting body called a perithecium. |

## P

| | |
|---|---|
| PARASITE | an organism that grows and feeds on another living organism; usually detrimental to the host organism. |
| PEDICEL | a small stalk-like structure. |
| PERENNIAL | an organism that grows several or more years. |
| PERIDIOLES | the structures in bird's nest fungi that contain the spores; dispersed as a whole structure; also structures in *Pisolithus tinctorius* that contain spores. |
| PERITHECIA | a microscopic flask-shaped fruiting body containing asci and ascospores; produced by some fungi. |
| PHENOL | a chemical that can cause color changes on some mushrooms. |
| PHOTOSYN-THESIS | a process performed by plants, algae and some bacteria that utilizes carbon dioxide and water to form an organic carbon source (glucose); primary producers. |
| PILEUS | the cap or top of a mushroom. |
| PLASMODIUM | the ephemeral vegetative stage of a slime mold life cycle. |
| POLYPORE | a group of fungi that are hard and woody with pores on the lower surface of the fruiting body. |
| PORE | an opening. |
| PUFFBALLS | a group of fungi with no stalk, but with a peridium containing the spores; spores are released through an ostiole or breaking of the peridium. |

## R

| | |
|---|---|
| RESUPINATE | flat, without a cap or bracket. |
| RETICULATE | netlike ridges. |
| RHIZINES | root-like structures for attachment on the lower surface of some lichen thalli. |
| RHIZOMORPHS | a persistent, hard, black, stringy modification of mycelium produced by some fungi. |

| | |
|---|---|
| RIDGED FUNGI | a group of fungi with fruiting bodies that have ridges on the lower surface; spores are produced in the ridged area. |
| RING | an annulus. |

## S

| | |
|---|---|
| SAPROBE | an organism that utilizes (feeds on) dead organic substrates and returns valuable nutrients to the soil; a decomposer. |
| SCABER | projecting tufted hairs; on stalks of some boletes. |
| SCIENTIFIC NAME | the Latin name used by biologists, consisting of two parts, the genus and the specific epithet. |
| SEXUAL REPRODUCTION | involves fusion of nuclei and subsequent reduction of genetic material by meiotic cell division; a form of reproduction that results in different genetic material being formed; the type of reproduction that happens on fruiting bodies of fungi. |
| SHELF | a bracket-shaped fruiting body. |
| SLIME MOLDS | organisms that have a plasmodium as the vegetative stage of the life cycle and form sporangia (fruiting bodies) containing spores. |
| SPHAEROCYSTS | round-shaped cells; occur in the cuticle or cap tissue of some mushrooms. |
| SPORES | a microscopic reproductive propagule that has the potential to germinate and grow into another organism. |
| STALK | supports the upper portion of a fruiting body; the stipe of a mushroom. |
| STINKHORNS | a group of fungi that have a stalk and an upper grooved portion containing spores in a slimy matrix. |
| STIPE | a stalk of mushrooms. |
| STRATIFIED | layered. |
| STRIATED | lined, or grooved. |
| STROMA | a modification of mycelium consisting of a firm, compact growth. |

## T

| | |
|---|---|
| THALLUS | the body of a lichen; crustose, foliose and fruticose forms. |
| TOADSTOOL | a term used by some individuals for a poisonous mushroom. |
| TOOTHED FUNGI | a group of fungi that have tooth-like structures on the lower surface of the cap. |

## U

**UMBO** — raised area, usually centrally located on the cap of a mushroom.

## V

**VEGETATIVE** — the portion of fungi not involved in sexual reproduction; occurs in the soil or substrate upon which the fungus is growing and usually is not visible; consists of hyphae/mycelium.

**VEIL** — a covering enclosing portions of some mushrooms when immature; partial veils cover the gills and universal veils enclose the entire mushroom.

**VOLVA** — a cup located below the soil from which some mushrooms emerge during development; it forms from a universal veil.

## W

**WHITE ROT** — a decomposition of wood resulting in a white appearance due to the simultaneous decay of lignin and cellulose.

# Common and Scientific Names of Plants

| Common Name | Scientific Name |
|---|---|
| Black Hills Spruce | *Picea glauca* (Moench) var. *densata* Bailey |
| Bog Birch | *Betula glandulosa* Michx. |
| Boxelder | *Acer negundo* L. |
| Bracken Fern | *Pteridium aquilinum* (L.) Kühn |
| Bur Oak | *Quercus macrocarpa* Michx. |
| Chokecherry | *Prunus virginiana* L. |
| Ironwood (=Hop Hornbeam) | *Ostrya virginiana* (Mill.) K. Koch |
| Mountain Juniper | *Juniperus scopulorum* Sarg. |
| Paper Birch | *Betula papyrifera* Marsh |
| Ponderosa pine | *Pinus ponderosa* Laws. |
| Rye | *Secale cereale* L. |
| Serviceberry | *Amelanchier alnifolia* Nutt. |
| Smooth Brome | *Bromus inermis* Leyss. |
| Trembling Aspen | *Populus tremuloides* Michx. |

# SELECTED REFERENCES

Alexopoulos, C.J., C.W. Mims and M. Blackwell. 1996. Introductory mycology. John Wiley & Sons, New York. 869 pp.

Arora, D. 1986. Mushrooms demystified, a comprehensive guide to the fleshy fungi. Ten Speed Press, Berkeley, California. 959 pp.

Barron, G. 1999. Mushrooms of northeast North America. Lone Pine Publishing, Renton, Washington. 336 pp.

Bessette, A. E., W. Roody and A. R. Bessette. 2000. North American boletes, a color guide to fleshy pored mushrooms. Syracuse University Press, Syracuse, New York. 396 pp.

Bessette, A. E., A. R. Bessette and D. Fischer. 1997. Mushrooms of northeastern North America. Syracuse University Press, Syracuse, New York. 582 pp.

Brodo, I., S. D. Sharnoff and S. Sharnoff. 2001. Lichens of North America. Yale University Press, New Haven, Conn. 795 pp.

Coker, W. C. and J. N. Couch. 1974. The gasteromycetes of the eastern United States and Canada. Dover Publications, New York. 201 pp.

Cummingham, G. H. 1979. The gasteromycetes of Australia and New Zealand. J. Cramer, Vaduz, Germany, 236 pp.

Dennis, R. W. G. 1977. British ascomycetes. J. Cramer, Germany. 585 pp.

Evenson, V. 1997. Mushrooms of Colorado and the southern Rocky Mountains. Denver Botanic Gardens and Westcliffe Publishers, Englewood, Colorado. 208 pp.

Farr, M.L. 1981. How to know the true slime molds. Wm. C. Brown Company Publishers, Dubuque, Iowa. 132 pp.

Gabel, A., E. A. Ebbert and K. Lovett. 2004. Macrofungi Collected from the Black Hills of South Dakota and Bear Lodge Mountains of Wyoming. American Midland Naturalist. (In Press).

Gilbertson, R. 1974. Fungi that decay ponderosa pine. University of Arizona Press, Tucson. 197 pp.

Groves, J. 1975. Edible and poisonous mushrooms of Canada. Canada Department of Agriculture, Ottawa, Ontario. 316 pp.

Hale, M. 1979. How to know the lichens. Wm. C. Brown Company Publishers, Dubuque, Iowa. 246 pp.

Horn, B., R. Kay and D. Abel. 1993. A guide to Kansas mushrooms. University Press of Kansas, Lawrence, Kansas. 297 pp.

Huffman, D., L. Tiffany and G. Knaphus. 1989. Mushrooms and other fungi of the midcontinental United States. Iowa State University Press, Ames, Iowa. 326 pp.

Kendrick, B. 2000. The fifth kingdom. Focus Publishing, Newburyport, Mass. 386 pp.

Lindsey, J. P. and R. Gilbertson. 1978. Basidiomycetes that decay aspen in North America. J. Cramer, Germany. 406 pp.

Martin, G. W. 1949. North American flora – fungi, myxomycetes. New York Botanical Garden, New York. 190 pp.

Marriott, H., D. Faber-Langendoen, A. McAdams, D. Stutzman and B. Burkhart. 1999. Black Hills community inventory report. The Nature Conservancy, Minneapolis, Minn. 175 pp.

Martin, G. W. , C. J. Alexopoulos and M. L. Farr. 1983. The genera of myxomycetes. University of Iowa Press, Iowa City, Iowa. 561 pp.

Miller, O. 1977. Mushrooms of North America. E.P. Dutton, New York. 368 pp.

Ostlie, W. R., R. E. Schneider, J. M. Aldrich, T. M. Faust, R. L. B. McKim and S. J. Chaplin. 2003. The status of biodiversity in the great plains. The Nature Conservancy, Minneapolis, Minn. http://www.greatplains.org/resource/biodiver/biostat/ecoregio.htm

Phillips, R. 1981. Mushrooms of North America. Little, Brown and Company, Boston, Mass., 319 pp.

Smith, A., H. Smith and N. Weber. 1979. How to know the gilled mushrooms. W. C. Brown, Dubuque, Iowa. 334 pp.

Smith, A., H. Smith and N. Weber. 1981. How to know the non-gilled mushrooms. W. C. Brown, Dubuque, Iowa. 324 pp.

Stamets, P. 1996. Psilocybin mushrooms of the world. Ten Speed Press, Berkeley, California. 245 pp.

Trappe, J. 1962. Fungus associates of ectotrophic mycorrhizae. Botanical Review 28: 538-606.

Tylutki, E. 1987. Mushrooms of Idaho and Pacific Northwest, vol 2., non-gilled Hymenomycetes. University of Idaho Press, Moscow, Idaho. 232 pp.

Van Bruggen, T. 1985. The vascular plants of South Dakota. Iowa State University Press, Ames, Iowa. 476 pp.

Vitt, D., J. Marsh and R. Bovey. 1988. Mosses, lichens and ferns. Lone Pine Publishing, Redmond, Washington. 296 pp.

Weber, N. S. 1995. A morel hunter's companion. Thunder Bay Press, Lansing, Michigan. 209 pp.

Wetmore, C. M. 1968. Lichens of the Black Hills of South Dakota and Wyoming. Stone Printing Company, Lansing, Michigan. 464 pp.

Ziller, W. 1974. The tree rusts of western Canada. Canadian Forestry Service Publication No. 1329, Victoria, British Columbia. 272 pp.

# Photograph and Illustration Credits

All photographs were taken by the authors except those listed below:

Kristie Lovett — Illustrations (Figures 9, 11, and 12) on pages 17, 19 and 20. The photograph of *Clitocybe nuda* on page 45, the top photograph of *Gloeophyllum sepiarium* on page 84, the photographs of *Pycnoporus cinnabarinus* on page 88, the photograph of *Coltrichia perennis* on page 93, the top photograph of *Phaeolus schweinitzii* on page 95, the photograph of *Cyathus striatus* on page 115, the photograph of *Cladonia* species on page 133 and the top photograph of *Fuligo septica* on page 140.

Mark Gabel — Photographs of *Scleroderma citrinum* on page 106 and *Pisolithus tinctorius* on page 105.

Larry Ebbert — Illustration of map on page 3 and habitat photograph of *Psathyrella candolleana* on page 67.

# County Locations of Featured Fungi

Following is a table showing the counties in the Black Hills area where each specimen has been collected from 1998-2002. The numerous collections from Lawrence and Pennington Counties reflect better collecting conditions due to higher moisture levels in the northern Black Hills. We collected from 95 sites throughout the Hills, but some sites were visited only once and others more frequently. Our collections do not indicate that the county/ counties listed are the only counties where a particular specimen could grow. They do indicate that our timing was correct for observing a particular fungus at a site. Documentation of future collections will probably extend the county range for many specimens. County locations for Group 6, Parasites and Group 7, Lichens are not listed because the survey was not designed to document their distribution.

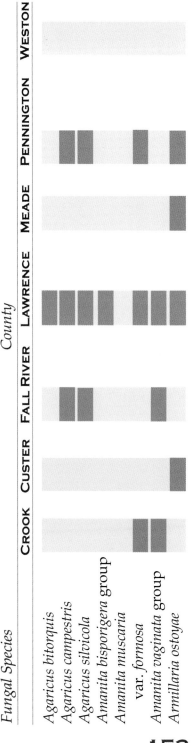

*Fungal Species*

| | | | County | | | | |
|---|---|---|---|---|---|---|---|
| | CROOK | CUSTER | FALL RIVER | LAWRENCE | MEADE | PENNINGTON | WESTON |
| *Agaricus bitorquis* | | | | | | | |
| *Agaricus campestris* | | | | ■ | | ■ | |
| *Agaricus silvicola* | | | ■ | ■ | | | |
| *Amanita bisporigera* group | | | ■ | ■ | | ■ | |
| *Amanita muscaria* var. *formosa* | | | | ■ | | | |
| *Amanita vaginata* group | ■ | | ■ | ■ | | ■ | |
| *Armillaria ostoyae* | | ■ | | ■ | ■ | ■ | |

153

## Fungal Species / County

| Fungal Species | Crook | Custer | Fall River | Lawrence | Meade | Pennington | Weston |
|---|---|---|---|---|---|---|---|
| Astraeus hygrometricus | ■ | | ■ | | ■ | | |
| Auricularia auricula | ■ | | | | | ■ | |
| Bjerkandera adusta | | | | ■ | | ■ | ■ |
| Boletus edulis | | | | ■ | | ■ | |
| Calvatia booniana | | | ■ | ■ | | | |
| Calvatia craniiformis | | | ■ | ■ | | | |
| Calvatia cyathiformis | | | | ■ | | | |
| Cerationmyxa fruticulosa | | | | ■ | | ■ | |
| Clavariadelphus truncatus | ■ | | | ■ | | ■ | |
| Clitocybe gibba | | ■ | | ■ | ■ | ■ | |
| Clitocybe nuda | | | | ■ | | ■ | |
| Coltrichia perennis | | | ■ | ■ | | ■ | |
| Coprinus comatus | | | | ■ | | ■ | |
| Coprinus lagopus | | | | ■ | | | |
| Coprinus micaceus | | | ■ | ■ | ■ | ■ | |
| Coprinus niveus | | | | ■ | | | |
| Coprinus quadrifidus | | | | ■ | | | |
| Cortinarius species | | ■ | | ■ | | ■ | |
| Crucibulum laeve | ■ | ■ | | ■ | | ■ | |
| Cryptoporus volvatus | ■ | | ■ | ■ | | | |
| Cyathus striatus | ■ | | | ■ | | ■ | |

154

*Fungal Species* / *County*

| Fungal Species | CROOK | CUSTER | FALL RIVER | LAWRENCE | MEADE | PENNINGTON | WESTON |
|---|---|---|---|---|---|---|---|
| Discina perlata | ■ | | | ■ | | ■ | |
| Fomes fomentarius | ■ | | | ■ | | | |
| Fomitopsis cajanderi | | ■ | ■ | ■ | ■ | ■ | |
| Fomitopsis pinicola | ■ | ■ | ■ | ■ | | | |
| Fuligo megaspora | | ■ | ■ | ■ | | ■ | |
| Fuligo septica | | | | ■ | ■ | | |
| Geastrum quadrifidum | ■ | | | ■ | | | |
| Geastrum saccatum | | | | ■ | | | |
| Geastrum triplex | | | | ■ | | | |
| Gloeophyllum sepiarium | | | | ■ | | ■ | |
| Gomphus clavatus | ■ | | | ■ | | ■ | |
| Gymnopilus sapineus | | | | ■ | | ■ | |
| Gymnopus dryophilus | | | ■ | ■ | | ■ | |
| Gyromitra gigas | | | | ■ | | ■ | |
| Gyromitra infula | | | | ■ | | | |
| Helvella ephippium | | | | ■ | | | |
| Hygrocybe conica | ■ | ■ | | ■ | | ■ | |
| Hygrophoropsis aurantiaca | ■ | ■ | | | | | |
| Hygrophorus chrysodon | | | ■ | ■ | | ■ | |
| Hygrophorus speciosus | | | ■ | ■ | | ■ | |
| Inocybe fastigiata | | | | | | ■ | ■ |

*Fungal Species* / County

| Fungal Species | Crook | Custer | Fall River | Lawrence | Meade | Pennington | Weston |
|---|---|---|---|---|---|---|---|
| Inonotus rheades | | | | ■ | | | ■ |
| Lactarius aquifluus | | | | ■ | | | |
| Lactarius mammosus | | | | ■ | | | |
| Leccinum insigne | | | | ■ | | ■ | |
| Leccinum snellii | | | | ■ | | ■ | |
| Lepiota naucina | | | | ■ | | ■ | |
| Leucopaxillus albissimus | | | ■ | ■ | | ■ | |
| Lycogala epidendrum | ■ | ■ | ■ | ■ | | | |
| Lycoperdon marginatum | | ■ | | ■ | | ■ | |
| Lycoperdon perlatum | | | | ■ | | ■ | |
| Lycoperdon pyriforme | | | | | | | |
| Morchella angusticeps | | | | ■ | | | |
| Morchella esculenta | | | ■ | ■ | ■ | ■ | |
| Mycena pura | | ■ | | | | | |
| Panaeolus campanulatus group | | | | ■ | | | |
| Peniophora rufa | ■ | | | ■ | | ■ | |
| Peziza badia | ■ | | | ■ | | ■ | |
| Peziza repanda | | | | ■ | | | |
| Phaeolus schweinitzii | | | | ■ | | ■ | |
| Phallus impudicus | | | ■ | ■ | | ■ | |

156

*Fungal Species* — County

| Fungal Species | Crook | Custer | Fall River | Lawrence | Meade | Pennington | Weston |
|---|---|---|---|---|---|---|---|
| *Phellinus tremulae* | ■ |  |  | ■ |  | ■ |  |
| *Pholiota vernalis* |  | ■ |  | ■ |  |  |  |
| *Physarum diderma* | ■ |  |  | ■ |  |  |  |
| *Pisolithus tinctorius* |  |  |  |  |  |  |  |
| *Pleurotus populinus/ostreatus* |  |  |  | ■ |  |  |  |
| *Pluteus cervinus* | ■ | ■ | ■ | ■ |  | ■ |  |
| *Polyporus arcularius* | ■ | ■ |  | ■ | ■ | ■ |  |
| *Polyporus badius* | ■ |  |  | ■ |  |  |  |
| *Polyporus cinnabarinus* | ■ | ■ | ■ | ■ | ■ | ■ |  |
| *Psathyrella candolleana* |  |  |  | ■ |  | ■ |  |
| *Psathyrella gracilis* |  |  |  | ■ |  | ■ |  |
| *Ramaria apiculata* | ■ |  |  | ■ |  |  |  |
| *Rhodotus palmatus* | ■ |  |  | ■ | ■ |  |  |
| *Russula albonigra* |  |  |  | ■ |  | ■ |  |
| *Russula alutacea group* |  |  |  | ■ |  | ■ |  |
| *Russula brevipes* | ■ | ■ | ■ | ■ |  | ■ |  |
| *Russula fragilis* |  |  |  | ■ |  |  |  |
| *Russula species* |  |  |  | ■ |  | ■ | ■ |
| *Sarcodon imbricatum* |  |  |  | ■ |  | ■ |  |
| *Scleroderma cepa* |  |  |  | ■ |  | ■ |  |

*Fungal Species* / *County*

| Fungal Species | CROOK | CUSTER | FALL RIVER | LAWRENCE | MEADE | PENNINGTON | WESTON |
|---|---|---|---|---|---|---|---|
| *Scleroderma citrinum* | ■ | | | ■ | | | |
| *Scutellinia scutellata* | ■ | | | ■ | | ■ | |
| *Sowerbyella rhenana* | | | | ■ | | | |
| *Stemonitis fusca* | ■ | | | ■ | | | |
| *Stropharia kauffmanii* | ■ | | | ■ | | ■ | |
| *Suillus granulatus* | ■ | | | ■ | ■ | ■ | |
| *Trametes hirsuta* | | ■ | ■ | ■ | | ■ | |
| *Trametes versicolor* | ■ | | | ■ | | ■ | |
| *Tremella mesenterica* | | | | ■ | | ■ | |
| *Tricholomopsis rutilans* | | | | ■ | | ■ | |
| *Xeromphalina campanella* | | | | ■ | | ■ | ■ |

# Index

## A

*Agaricus bitorquis* (Quél.) Sacc., 65
*Agaricus campestris* Fr., 65
*Agaricus silvicola* (Vitt.) Peck, 30, 64
*Aleuria rhenana* Fuckel, 120
*Amanita bisporigera* group, 16, 17, 27, 34
*Amanita muscaria* var. *formosa* (Pers. ex Fr.) Bertillon, 27, 32
*Amanita vaginata* group, 27, 33
*Armillaria mellea* (Vahl. ex Fr.) P. Kumm., 44
*Armillaria ostoyae* (Romagn.) Herink, 28, 43
Aspen Bolete, 76, 79
*Astraeus hygrometricus* (Pers.) Morgan, 104, 110
*Auricularia auricula* (Hook.) Underw., 83, 99

## B

Bell-Shaped Panaeolus, 29, 59
Bird's Nest Fungi, 23
*Bjerkandera adusta* (Willd. ex Fr.) Karst., 82, 90
Black Knot, 127
Black-Leg, 82, 96
Black Morel, 117, 122
Blackening Russula, 29, 55
Black-Spored Mushrooms, 59
Blewit, 28, 45
Boggy Cortinarius, 73
Boletes, 20, 77
*Boletus edulis* Bull. ex Fr., 76, 80
Brittle, White to Yellow-Spored Mushrooms, 49
Brown Bolete, 76, 78
Brown Squirrel Ears, 83, 99
Burnt Sugar Mushroom, 29, 49

## C

*Calvatia booniana* Smith, 104, 108
*Calvatia craniiformis* (Bosc.) Morg., 109
*Calvatia cyathiformis* (Bosc.) Morg., 109
*Cantharellus*, 38
*Ceratiomyxa fruticulosa* (Müll.) Macbr., 137
Chanterelles, 38
Chocolate Tubes, 138
*Cladonia* species, 133
*Clavariadelphus ligula* (Fr.) Donk, 103
*Clavariadelphus truncatus* (Quél.) Donk, 22, 23, 83, 103
*Claviceps purpurea* (Fr.) Tul., 130
*Clitocybe glaucocana* (Bres.) H. E. Bigelow, 45, 46
*Clitocybe gibba* (Fr.) P. Kumm., 16, 18, 28, 36, 46
*Clitocybe nuda* (Bull. ex Fr.) H. E. Bigelow & Sm., 28, 45, 46
Club Fungi, 21, 102
*Collybia acervata* (Fr.) P. Kumm., 48
*Collybia dryophila* (Bull. ex Fr.) P. Kumm., 48
*Coltricia perennis* (Fr.) Murray, 82, 93
Common Bird's Nest Fungus, 105, 115
Common Coral, 83, 102
Common or Boring Gymnopilus, 30, 70
Common Inky Cap, 62
Common Spiny Puffball, 104, 111
Common Stinkhorn, 104, 107
Common Woodland Psathyrella, 30, 67
Conical Waxy Cap, 28, 39
Convoluted Cup Fungus, 117, 118
*Coprinus atramentarius* (Bull. ex Fr.) Fr., 62
*Coprinus comatus* (Müll. ex Fr.) S. F. Gray, 30, 60

**159**

*Coprinus lagopus* (Fr.) Fr., 63
*Coprinus micaceus*
  (Bull. ex Fr.) Fr., 30, 61
*Coprinus niveus*
  (Pers. ex Fr.) Fr., 63
*Coprinus quadrifidus* Peck, 63
Coral Fungi, 21, 102
*Cortinarius* species,
  16, 17, 30, 70, 73
*Crucibulum laeve* (Huds.) Kamb.,
  23, 24, 105, 115
Cryptic Globe Fungus, 81, 85
*Cryptoporus volvatus*
  (Peck) Shear, 81, 85
Cup Fungi, 24, 118
*Cyathis striatus*
  (Huds.) Pers., 105, 114

### D

Destroying Angel, 27, 34
*Dibotryon morbosum*
  (Schwein.) Theiss. & Syd., 127
*Discina perlata* (Fr.) Fr., 117, 118
Dye Polypore, 82, 95

### E

Earth Stars, 23
Eyelash Cup, 117, 121
*Endocronartium harknessii*
  (J. P. Moore) Y. Hirat., 131
Ergot, 130
Ergotism, 130

### F

Fairy Stool, 82, 93
False Chanterelle, 28, 38
False Orange Peel Cup, 117, 120
False Tinder Polypore, 81, 87
Fawn Mushroom, 29, 57
Fly Agaric, 27, 32
*Fomes fomentarius*
  (L. ex Fr.) Kickx, 87
*Fomitopsis pinicola* (Swartz ex Fr.)
  Karst., 21, 22, 81, 85, 86
*Fomitopsis cajanderi* (Karst.)
  Kotlaba & Pouzar, 86

Fragile Russula, 29, 54
Fringed Polypore, 82, 94
*Fuligo megaspora* Sturgis, 140
*Fuligo septica* (L.) Wiggers, 139
Funnel Cap, 28, 36

### G

*Geastrum quadrifidum* Pers.
  23, 104, 109
*Geastrum saccatum* Fr., 110
*Geastrum triplex* Jung., 110
Giant Western Puffball, 104, 108
*Gloeophyllum sepiarium*
  (Fr.) Karst., 81, 84
*Gomphus clavatus*
  (Fr.) Gray, 22, 83, 100
Graceful Psathyrella, 30, 68
Granulated Slippery Jack, 76, 77
Gray Dunce Caps, 28, 42
Gray-Saddled False Morel,
  117, 126
Green Fairy Cups, 133
Green Lobster, 129
Grisette, 27, 33
*Gymnopilus sapineus*
  (Fr.) Maire, 30, 70
*Gymnopus acervatus*
  (Fr.) Murrill, 48
*Gymnopus dryophilus*
  (Bull. ex Fr.) Murrill, 28, 48
*Gyromitra gigas* (Krombh.)
  Quél., 24, 25, 117, 124
*Gyromitra infula*
  (Schaeff. ex Fr.) Quél., 117, 125

### H

Hairy Turkey Tail, 82, 92
Hard Puffball, 104, 106
*Helvella ephippium* Lév.,
  25, 117, 126
Herds of White Crust, 138
Honey Mushroom, 28, 43
Horse Hoof Fungus, 104, 105
*Hygrophoropsis aurantiaca*
  (Wulf. ex Fr.) Maire, 28, 38

*Hygrocybe conica* (Scop.ex Fr.)
  P.Kumm., 28, 39
*Hygrophorous chrysodon*
  (Fr.) Fr., 28, 40, 41
*Hygrophorous speciosus*
  Peck, 16, 18, 28, 39, 40
*Hypomyces lactifluorum*
  (Schw. ex Fr.) Tul., 128
*Hypomyces luteovirens*
  (Fr.: Fr.) Tul., 129

**I**
*Inocybe fastigiata*
  (Schaeff. ex Fr.) Quél., 30, 72
*Inocybe geophylla*
  (Sow. ex Fr.) P. Kumm., 73
*Inonotus rheades* (Pers.)
  Bond. et Sing., 81, 89

**J**
Jack-O-Lantern Mushroom, 38
Jelly Fungi, 21, 98

**K**
King Bolete, 76, 80
*Kuehneromyces vernalis* (Peck)
  Singer & Smith, 71

**L**
*Lactarius aquifluus* Peck, 29, 49
*Lactarius hibbardae* Peck, 51
*Lactarius mammosus*
  (Weinm. ex Fr.) Fr., 29, 51
*Leccinum insigne* Smith,
  Thiers & Watling, 21, 76, 79
*Leccinum snellii* Smith,
  Thiers & Watling, 76, 78
*Lepiota naucina*
  (Fr.) P. Kumm., 27, 31
*Lepista nuda* (Bull. ex Fr.) Cke., 45
*Leucoagaricus naucinus*
  (Fr.) Singer, 31
*Leucopaxillus albissimus*
  (Pk.) Singer, 28, 46
Lichens, 26, 132
Lobster Mushroom, 128
*Lycogala epidendrum* (L.) Fr., 140

*Lycoperdon marginatum*
  Vitt., 104, 112
*Lycoperdon perlatum*
  Pers., 23, 104, 111
*Lycoperdon pyriforme*
  Pers., 104, 113

**M**
Meadow Mushroom, 65
*Melanoleuca* species, 46
Mica Cap, 30, 61
Milky Mushroom, 29, 51
*Morchella angusticeps*
  Peck, 25, 117, 122
*Morchella esculenta* Fr., 117, 123
Mushrooms, 16, 31
*Mycena* species, 43
*Mycena pura* (Pers. ex Fr.)
  P. Kumm., 28, 42
*Mycena purpureofusca*
  (Pk.) Sacc., 42

**N**
Netted Mushroom, 29, 58
Non-Brittle–White-Spored
  Mushrooms, 31

**O**
Oak Gymnopus, 28, 48
Old Man's Beard, 134
*Omphalotus* species, 38
Orange Brown, Rusty Brown to
  Brown-Spored Mushrooms, 70
Orange-Saddled
  False Morel, 117, 125
Oyster Mushroom, 27, 35

**P**
*Panaeolus campanulatus*
  group, 29, 59
Parasitic Fungi, 25, 127
*Paxillus involutus* (Fr.) Fr., 49
Pear-Shaped Puffball, 104, 113
Peeling Puffball, 104, 112
*Peltigera aphthosa* (L.) Willd., 136
*Peltigera canina* (L.) Willd., 136
*Peniophora rufa* (Fr.) Boud., 83, 101

*Peziza badia* Pers. ex Mérat, 120
*Peziza repanda* Pers.,
   24, 25, 117, 119
*Phaeolus schweinitzii*
   (Fr.) Pat., 82, 95
*Phallus impudicus* Pers.,
   23, 24, 104, 107
*Phellinus tremulae* (Bond.)
   Bond. et Boriss., 81, 87
*Pholiota vernalis* (Peck)
   Smith & Hesler, 30, 71
*Physarum diderma* Rost, 138
Pig's Ears, 83, 100
Pine Gall Rust, 131
Pine Waxy Cap, 28, 40
Pink and Tan Bubbles, 140
Pink-Spored Mushrooms, 57
*Pisolithus tinctorius* Pers., 104, 105
*Pleurotus ostreatus* (Jacq.: Fr.)
   P. Kumm., 27, 35
*Pleurotus populinus*
   Hilber & Miller, 27, 35
Plums and Custard, 28, 47
*Pluteus admirabilis* Peck, 58
*Pluteus cervinus*
   (Fr.) P. Kumm., 29, 57
Polypores, 21, 84
*Polyporus arcularius*
   Batsch ex Fr., 82, 94
*Polyporus badius* (Pers. ex S. F.
   Gray) Schw., 82, 96
*Psathyrella candolleana*
   (Fr.) Maire, 30, 67
*Psathyrella gracilis*
   (Fr.) Quél., 30, 68
Puffballs, 23
Purple Brown-Spored
   Mushrooms, 64
*Pycnoporus cinnabarinus*
   (Jacq.: Fr.) Karst., 81, 88

## R
*Ramaria apiculata* (Fr.) Donk
   22, 83, 102
*Ramaria stricta* (Fr.) Quél., 102
Red-Belted Conk, 81, 86

Red Polypore, 81, 88
Resupinate Fungi, 21, 101
*Rhodotus palmatus*
   (Bull. ex Fr.) Maire, 29, 58
Ridged Fungi, 21, 100
Rosy Conk, 86
Rosy-Red Russulas, 29, 52
*Russula albonigra*
   (Krombh.) Fr., 29, 55
*Russula alutacea* group, 29, 53
*Russula brevipes* Peck, 29, 56
*Russula emetica* group, 52
*Russula fragilis*
   (Pers. ex Fr.) Fr., 29, 54
*Russula* species, 29, 52
Rusty-Gilled Polypore, 81, 84
Rusty Polypore, 81, 89

## S
*Sarcodon imbricatum*
   (L. ex Fr.) Karst., 21, 22, 82, 97
Scaly Inky Cap, 63
Scaly Stropharia, 30, 66
Scaly-Toothed Fungus, 82, 97
Scrambled Egg Slime, 139
*Scleroderma cepa* Pers., 107
*Scleroderma citrinum* Pers., 104, 106
*Scutellinia scutellata* (L.: St.-Amans)
   Lambotte, 117, 121
Shaggy Mane, 30, 60
Shoestring Fungus, 44
Short-Stemmed Russula, 29, 56
Silky Little Brown
   Mushroom, 30, 72
Slime Molds, 26, 137
Slimy Cortinarius, 74
Smoky Polypore, 82, 90
Smooth Cup Fungus, 117, 119
Smooth Parasol, 27, 31
Snowbank False Morel, 117, 124
*Sowerbyella rhenana*
   (Fuckel) Moravec, 117, 120
Stalked Earth Star, 104, 109
*Stemonitis fusca* Roth, 138
Stinkhorns, 23

Striated Bird's Nest Fungus
   105, 114
*Stropharia kauffmanii* Smith, 30, 66
*Suillus granulatus*
   (Fr.) Kuntze, 20, 76, 77
Sunburst Orange Lichen, 135

**T**
Tan Morel, 117, 123
Tinder Polypore, 87
Tiny Trumpets, 28, 37
Toothed Fungi, 21, 97
*Trametes hirsuta*
   (Wulf.: Fr.) Pilát, 82, 92
*Trametes versicolor*
   (L. ex Fr.) Pilát, 82, 91
*Tremella mesenterica*
   (Gray) Pers., 21, 22, 82, 98
*Tricholoma*, 46
*Tricholomopsis decora*
   (Fr.) Singer, 47
*Tricholomopsis platyphylla*
   (Pers. ex Fr.) Singer, 47
*Tricholomopsis rutilans*
   (Schaeff.: Fr.) Singer, 28, 47
Truncate Club Coral, 83, 103
Turkey Tails, 82, 91

**U**
*Usnea cavernosa* Tuck., 134
*Usnea hirta* (L.) Wigg, 134

**V**
Variable Russula, 29, 53
Veined Dog Lichen, 136
Vernal Changing Pholiota, 30, 71

**W**
Warts on Wood, 83, 101
Water Measurer, 104, 110
White Coral Slime, 137
White Leucopax, 28, 46
Witch's Butter, 82, 98
Woodland Agaricus, 30, 64
Wooly Inky Cap, 63

**X**
*Xanthoria fallax* (Hepp.) Arn, 135
*Xanthoria hasseana* Räsänen, 135
*Xanthoria montana*
   L. Lindblom, 135
*Xanthoria polycarpa*
   (Ehrh.) Oliv., 135
*Xeromphalina campanella* (Bat.
   ex Fr.) Kühner & Maire, 28, 37
*Xeromphalina cauticinalis*
   (Fr.) Kühner & Maire, 37

**Y**
Yellow Waxy Cap, 28, 41

# Biographies

**AUTHOR:** Audrey Gabel received her M. S. in mycology/plant pathology and Ph.D. in mycology/botany from Iowa State University. She taught biology at Black Hills State University until her retirement in 2002. Former research has focused on fungal plant pathogens and she has published in Mycologia, Mycotaxon and other professional journals. As Professor Emeritus she remains active in mycological research.

**COAUTHOR:** Elaine Ebbert is a graduate of Black Hills State University with a B. S. in Biology. While a student she was active in research and has presented posters and papers in the state and nation. Currently, she works with the Black Hills Program of The Nature Conservancy. Her primary interests include mycology, botany and land conservation.

**ART DIRECTOR:** Linn Nelson received her BFA in design from the University of South Dakota and her MFA from Southern Illinois University. She has worked as a designer and creative director in marketing and advertising. Currently she is a professor of graphics at Black Hills State University. Ms. Nelson has received several awards for her designs from the South Dakota Advertising Federation.

**PHOTOGRAPHIC EDITOR:** Steve Babbitt received his BFA and MFA in photography from The San Francisco Art Institute. Currently he is a professor of photography at Black Hills State University. His photographs can be found in the collections of The Bibliotheque Nationale in Paris, The Getty Museum Library in California, The San Francisco Art Institute and the Sioux Falls Civic Fine Arts Center.

**BUSINESS MANAGER:** Daniel O. Farrington holds a DVM from Colorado State University and a Ph.D. in Veterinary Microbiology and Preventive Medicine from Iowa State University. He has been a veterinary practitioner and for over 20 years he worked as an animal health research director in the pharmaceutical industry. He is currently Vice President for Academic Affairs at Black Hills State University.

## BLACK HILLS STATE UNIVERSITY PRESS